Stanley Smyth Flower

Notes on a collection of Reptiles and Batrachians made in the

Malay Peninsula in 1895-96

Stanley Smyth Flower

Notes on a collection of Reptiles and Batrachians made in the Malay Peninsula in 1895-96

ISBN/EAN: 9783741115189

Manufactured in Europe, USA, Canada, Australia, Japa

Cover: Foto ©Klaus-Uwe Gerhardt /pixelio.de

Manufactured and distributed by brebook publishing software (www.brebook.com)

Stanley Smyth Flower

Notes on a collection of Reptiles and Batrachians made in the Malay Peninsula in 1895-96

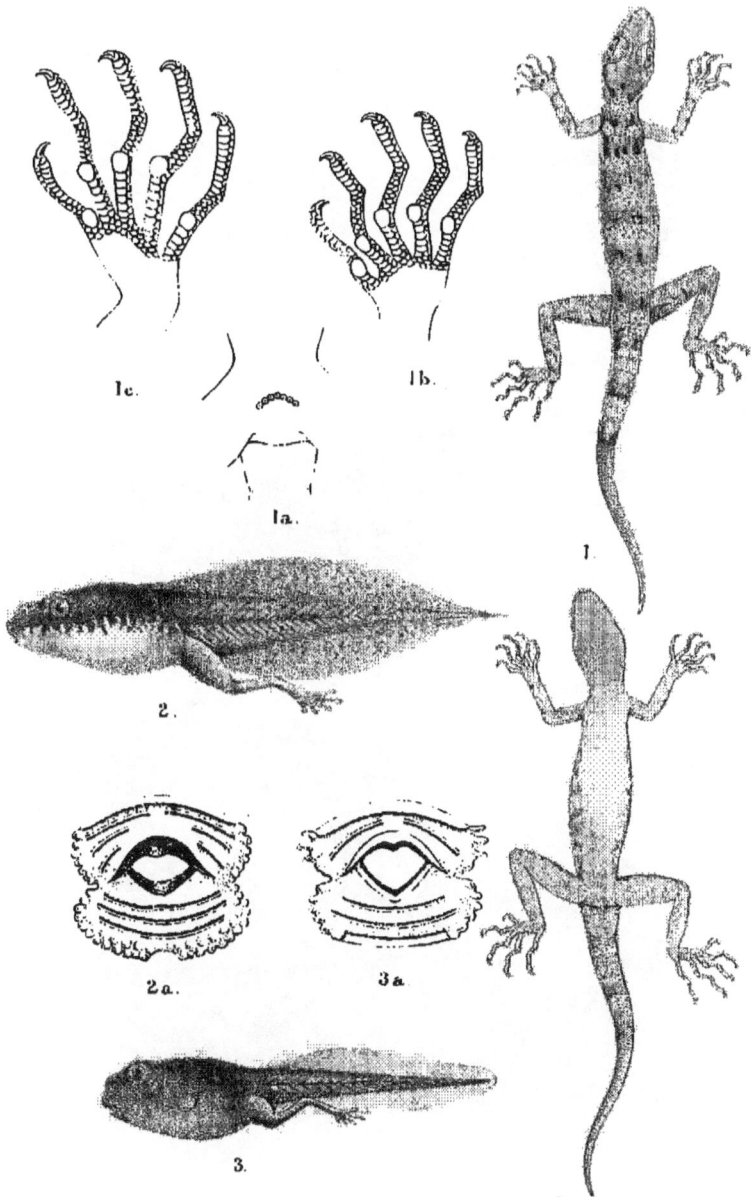

1c. 1b. 1a. 1. 2. 2a. 3a. 3.

S.S.F.& J.Green del.et.lith.

Mintern Bros. Chromo.

MALAY REPTILES AND BATRACHIANS.

1.*Gonatodes penangensis*. 2.*Rhacophorus leucomystax*.
3.*Bufo melanostictus*.

3a.

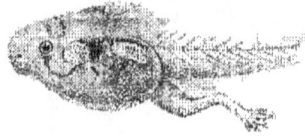

3.

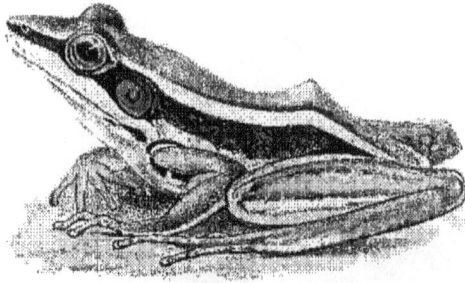

2.

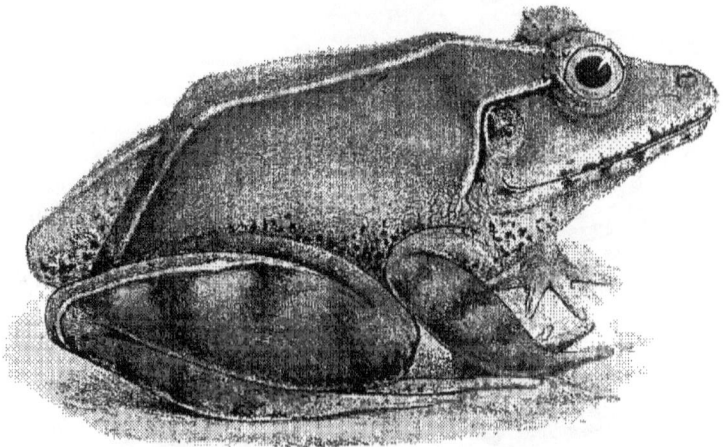

1.

S.S.F.del.

Mintern Bros. Chromo.

MALAY BATRACHIANS.

1 Rana macrodon, 2.Rana erythræa, 3.Rana labialis.

S.S.F. del.

Mintern Bros. Chromo.

MALAY BATRACHIANS.
Rana luctuosa.

1. Notes on a Collection of Reptiles and Batrachians made
   in the Malay Peninsula in 1895–96; with a List of the
   Species recorded from that Region.   By Stanley Smyth
   Flower, 5th Fusiliers.[1]

[Received October 15, 1896.]

(Plates XLIV.–XLVI.)

Since Dr. Cantor published his ' Catalogue of Reptiles inhabiting
the Malayan Peninsula and Islands' in 1847, no general list has
appeared: in his Catalogue mention is made of 106 species of
Reptiles and Batrachians; in this paper 210 species are listed.   Our
knowledge of the herpetological fauna of Malaya since Cantor's
time has been added to principally in two valuable papers by
Stoliczka in the Journal of the Asiatic Society of Bengal (1870,
vol. xxxix. part ii. pp. 134–228, and 1873, vol. xlii. part ii. pp. 111–
126), and by collections received in the British Museum from

[1] Communicated by the President.

Dr. N. B. Dennys, Mr. D. F. A. Hervey, Mr. H. N. Ridley, Mr. L. Wray, etc.   The specimens hitherto received have nearly all been collected in the more settled localities of the Peninsula, *i. e.* Penang, Province Wellesley, Perak, Malacca, and Singapore : the States of Kedah, Kelantan, Tringanu, Pahang, Johore, etc., are practically unexplored, so that it is probable that many additions are still to be made to the number of species of Reptiles and Batrachians from Malaya.   There are Museums at Taiping, Kuala Lumpur, and Singapore: the collection in the latter place I have had some opportunity for examining, but want of time did not enable me to do so as thoroughly as I could have wished; the Taiping Museum I have only paid a short visit to; the Kuala Lumpur one I have not seen : there is also a large private collection of Snakes at the Prye Estate, Province Wellesley, and probably collections at other places.   When all these have been thoroughly examined, we shall have a better knowledge of the relative abundance, localities, and varieties of the different species.

I have to acknowledge my sense of obligation to Mr. G. A. Boulenger for his most kind and useful advice to me both before starting to the East and in working out my collection on returning home; also to Mr. J. C. Somerville, 5th Fusiliers, and Mr. H. N. Ridley, Superintendent of the Botanical Gardens, Singapore, for assistance in collecting; I am also indebted to Mr. A. H. B. Dennys, of Penang, for a collection of Snakes made a few years ago in the Province Wellesley, and to Commissioner of Police Mitchell, of Kedah, for specimens from Kulim and neighbourhood.

In this paper the following species are recorded from the Malay Peninsula and adjacent islands for, I believe, the first time :— *Gonatodes penangensis*, sp. n., *Minnetozoon floweri*, Blgr., *Lepidodactylus ceylonensis*, Blgr., *Mabuia novemcarinata*, Anders., *Calamaria pavimentata*, Dum. & Bibr., *Rana luctuosa*, Peters, *Ixalus pictus*, Peters, and *Calophrynus pleurostigma*, Tschudi ; the variations of *Rana macrodon* are discussed ; and the tadpoles of four species are described.

The classification and nomenclature are according to the British Museum Catalogues of the Reptilia and Batrachia, to which valuable works I must refer for the complete synonymy of the various species mentioned ; I have only quoted the names under which they appear in the standard works referred to.

Order CHELONIA.

Suborder ATHECÆ.

Family SPHARGIDÆ.

1. DERMOCHELYS CORIACEA, L.

Dermochelys coriacea, Boul. Cat. Chel. etc. p. 10 (skull fig. p. 9).
There is a large specimen, unlabelled, in the Raffles Museum, Singapore, supposed to have been caught in the neighbourhood.

Hab. Tropical seas ; sometimes occurs in the temperate seas.

Suborder THECOPHORA.

Superfamily CRYPTODIRA.

Family TESTUDINIDÆ.

2. CALLAGUR PICTA, Gray.

Emys trivittata, Cantor, p. 4.
Tetraonyx affinis, part., Cantor, p. 6.
Batagur affinis, Günther, Rept. Brit. Ind. p. 40, pl. iii. fig. C.
Callagur picta, Boul. Cat. Chel. etc. p. 60.

There is a specimen in the British Museum from Penang through Cantor ; from his account this species appears not to be numerous, inhabiting the coasts, rivers, and ponds of Malaya.
Hab. Malay Peninsula and Borneo.

3. BATAGUR BASKA, Gray.

Tetraonyx affinis, part., Cantor, p. 6.
Batagur baska, Günther, Rept. Brit. Ind. p. 37, pl. iii. fig. B ; Boul. Cat. Chel. etc. p. 61 (skull fig. p. 62).

There is a specimen in the British Museum from the coast of Penang through Cantor.
Hab. Bengal, Burma, and Malay Peninsula.

4. HARDELLA THUROI, Gray.

Emys thurgi, Günther, Rept. Brit. Ind. p. 24.
Hardella thurgi, Boul. Cat. Chel. etc. p. 63 (skull fig. p. 64, and shell fig. p. 65).

Günther (R. B. I. p. 25) says that according to Cantor this species is found in Penang; but I do not find it mentioned in Cantor's Catalogue.
Hab. India (Indus, Ganges and tributaries), and perhaps Malay Peninsula.

5. BELLIA CRASSICOLLIS, Gray.

Emys crassicollis, Cantor, p. 3 ; Günther, Rept. Brit. Ind. p. 28, pl. iv. fig. E.
Bellia crassicollis, Boul. Cat. Chel. etc. p. 98 (skull fig. p. 98, and shell fig. p. 99).

Cantor says this species is numerous in the rivulets and ponds in the valleys of Penang and the Malay Peninsula. Stoliczka (J. A. S. B. 1870, vol. xxxix. part ii. p. 227) found it common in the small freshwater streams of Penang.
Hab. Tenasserim, Siam, Malay Peninsula, Sumatra, and Borneo.

6. CYCLEMYS PLATYNOTA, Gray.

Notochelys platynota, Günther, Rept. Brit. Ind. p. 17.
Cyclemys platynota, Boul. Cat. Chel. etc. p. 130.

There are five specimens in the British Museum from Singapore collected by Mr. A. R. Wallace. Cantor (p. 3) says that *Emys platynota* inhabits the valleys of the Malay Peninsula and Penang, but is apparently not numerous; however, Günther (R. B. I. p. 18) remarks, " this was certainly an incorrect determination, as is evident from his description" : I have not made out to what species Cantor's Penang Tortoise belongs.

Hab. Mergui, Malay Peninsula, Sumatra, and Borneo.

7. CYCLEMYS DHOR, Gray.

Cyclemys oldhamii, Günther, Rept. Brit. Ind. p. 15, pl. v. fig. B.
Cyclemys dhor, Boul. Cat. Chel. etc. p. 131.

It is stated in several works that this tortoise occurs in the Malay Peninsula; Dr. Gray (Cat. Shield Rept. 1855, p. 43) says that three young tortoises from Penang, described by Cantor (p. 6) as *Tetraonyx affinis,* were probably the young of this species; these specimens are now considered to belong to *Callagur picta* and *Batagur baska.*

Hab. Northern India, Burma, Siam, Camboja, Malay Peninsula and Archipelago.

8. CYCLEMYS AMBOINENSIS, Daud.

Cistudo amboinensis, Cantor, p. 5.
Cuora amboinensis, Günther, Rept. Brit. Ind. p. 12, pl. iv. figs. A, B.
Cyclemys amboinensis, Boul. Cat. Chel. etc. p. 133 (skull fig. p. 128, and shell fig. p. 129).

' There are specimens in the British Museum from Malacca and Singapore. Cantor says, " This species appears to be numerous in the valleys, in ponds, rivulets and paddy fields, Malayan Peninsula and Singapore.'' Mr. Ridley informed me he had found it plentiful at Malacca. I found two specimens near the Ayer Etam road in Penang: the length of carapace of the larger was 198 mm.

Hab. Burma, Siam, Malay Peninsula and Archipelago, extending eastward to the Moluccas.

9. GEOEMYDA SPINOSA, Gray.

Geoemyda spinosa, part., Cantor, p. 2.
Geoemyda spinosa, Günther, Rept. Brit. Ind. p. 18; Boul. Cat. Chel. etc. p. 137.

There are specimens in the British Museum from Penang (Cantor) and from Singapore (A. R. Wallace). Mr. Ridley has found this species on Bukit Timah, Singapore. In January 1896 I found two specimens in the water, in streams on the south side of Bukit Timah, Singapore; the length of carapace of the larger was 186 mm. In captivity these tortoises spent nearly all their time in the water; they fed daily, eating for their size large quantities of fruit, preferring pineapple, but also taking orange, banana, raisins, lettuce, etc. On the 14th of May one laid an oblong, equal-ended, white egg. These specimens are now alive in the Society's Gardens (August 1896).

Hab. Tenasserim, Malay Peninsula, Sumatra, and Borneo.

10. TESTUDO EMYS, Schl. & Müll.

Geoemyda spinosa, part., Cantor, p. 2.
Manouria emys, Günther, Rept. Brit. Ind. p. 10.
Testudo emys, Boul. Cat. Chel. etc. p. 158 (skull fig. p. 150).

Cantor records this species from the Great Hill of Penang; there are two Penang specimens from him in the British Museum. In March 1896 Mr. Ridley caught a fine female specimen in the Dindings, and brought it alive to the Botanical Gardens, Singapore; the length of carapace was about 520 mm.

Hab. Assam, Burma, Siam, Malay Peninsula, Sumatra, and Borneo.

Family CHELONIDÆ.

11. CHELONE MYDAS, L.

Chelonia virgata, Cantor, p. 11; Günther, Rept. Brit. Ind. p. 53.
Chelone mydas, Boul. Cat. Chel. etc. p. 180.

Cantor says, "This species is at all seasons plentifully taken in fishing-stakes in the Straits of Malacca."
Hab. Tropical and subtropical seas.

12. CHELONE IMBRICATA, L.

Chelonia imbricata, Cantor, p. 13.
Caretta squamata, Günther, Rept. Brit. Ind. p. 54.
Chelone imbricata, Boul. Cat. Chel. etc. p. 183 (skull fig. p. 181); Boul. Fauna Ind., Rept. p. 49 (young fig.).

Cantor mentions this species as inhabiting Malayan seas. Dr. Hanitsch showed me a live specimen caught near Singapore early in 1896.
Hab. Tropical and subtropical seas.

13. THALASSOCHELYS CARETTA, L.

Chelonia olivacea, Cantor, p. 13.
Caouana olivacea, Günther, Rept. Brit. Ind. p. 52.
Thalassochelys caretta, Boul. Cat. Chel. etc. p. 184.

Cantor says, "This species is at Penang of rare occurrence." A specimen (now preserved in the Raffles Museum) was caught near Singapore in Jan. 1896 : length of carapace 700 mm.
Hab. Tropical and subtropical seas.

Superfamily TRIONYCHOIDEA.

Family TRIONYCHIDÆ.

14. TRIONYX SUBPLANUS, Geoffr.

Trionyx subplanus, Günther, Rept. Brit. Ind. p. 49.
Trionyx güntheri, Günther, Rept. Brit. Ind. p. 49, pl. iv. fig. 4.
Trionyx subplanus, Boul. Cat. Chel. etc. p. 246 (skull fig. p. 247).

There are specimens in the British Museum from Penang from

Dr. Cantor, and from Singapore from Gen. Hardwicke and Mr. A.
R. Wallace.
Hab. Mergui, Malay Peninsula, Sumatra, Borneo, and Java.

15. TRIONYX HURUM, Gray.

Gymnopus gangeticus, Cantor, p. 8.
Trionyx gangeticus, Günther, Rept. Brit. Ind. p. 47.
Trionyx hurum, Boul. Cat. Chel. etc. p. 249 ; Boul. Fauna Ind.
Rept. p. 13 (young figured).

Cantor says this species inhabits the rivers and sea-coasts of
Penang and the Malay Peninsula, but that it is not numerous.
Hab. Ganges and Malay Peninsula.

16. TRIONYX PHAYRII, Theob.

Trionyx phayrii, Boul. Cat. Chel. etc. p. 251 (skull fig. p. 252).
Anderson (J. A. S. B. 1871, p. 30) mentions a specimen of
Trionyx phayrei, Theobald, in the Calcutta Museum from Penang.
Hab. Burma, Malay Peninsula ; Java, Borneo.

17. TRIONYX CARTILAGINEUS, Bodd.

Gymnopus cartilaginea, Cantor, p. 9.
Trionyx ornatus, Günther, Rept. Brit. Ind. p. 48, pl. iv. fig. B.
Trionyx cartilagineus, Boul. Cat. Chel. etc. p. 253 (skull figured).

There is in the British Museum a Penang specimen from Cantor ;
he says, " This species is numerous in rivers and ponds, Malayan
Peninsula and Pinang."
Hab. Burma, Siam, Camboja, Malay Peninsula, Sumatra, Borneo,
and Java.

18. PELOCHELYS CANTORIS, Gray.

Gymnopus indicus, Cantor, p. 10.
Chitra indica, Günther, Rept. Brit. Ind. p. 50, pl. vi. fig. C.
Pelochelys cantoris, Boul. Cat. Chel. etc. p. 263 (skull fig. p. 262).

The type specimen is in the British Museum from Penang
from Cantor ; he says it inhabits the estuaries and sea-coasts of
Penang and the Malay Peninsula.
Hab. Ganges, Burma, China, Malay Peninsula, Borneo, and
Philippines.

Order EMYDOSAURIA.

Family CROCODILIDÆ.

1. TOMISTOMA SCHLEGELI, S. Müll.

Tomistoma schlegelii, Boul. Cat. Chel. etc. p. 276 ; id. P. Z. S.
1896, p. 628.

There is a specimen in the British Museum from Pulo Tiga,
Perak river, given by Mr. Wray in 1896 ; and there is a skull

from the Perak river in the Taiping Museum. I have heard of a third specimen from Perak and also of its being found in Pahang. *Hab.* Malay Peninsula, Sumatra, and Borneo.

2. CROCODILUS POROSUS, Schn.

Crocodilus porosus, Cantor, p. 16 ; Günther, Rept. Brit. Ind. p. 62.
Crocodilus pondicerianus, Günther, Rept. Brit. Ind. p. 62, pl. vii.
Crocodilus porosus, Boul. Cat. Chel. etc. p. 284.

Cantor says this species is exceedingly numerous in the Malay Peninsula, Penang, and Singapore. Stoliczka (J. A. S. B. 1873, p. 113) found it in the collection he got from Penang and Province Wellesley. There are several stuffed specimens and a large skull of this species in the Raffles Museum ; one shot by Mr. Owen at Serangoon, Singapore, measures 4·7 metres in total length. In April 1895 I saw many Crocodiles of this species on the Kedah river, between Kuala Kedah and Kota Star, when lying up on the mud-banks under the trees ; their markings and vivid yellow and black colouring render them hard to see in the chequered light coming through the foliage. I have also seen this species on the Pandan river, Singapore. It is probably found in every suitable locality in Malaya.

Hab. India, Ceylon, Burma, Siam, Southern China, Malay Peninsula and Archipelago, New Guinea, North Australia, Solomon and Fiji Islands.

3. CROCODILUS PALUSTRIS, Less.

Crocodilus vulgaris, Cantor, p. 15.
Crocodilus palustris, Günther, Rept. Brit. Ind. p. 61, pl. vii. fig. A ; Boul. Cat. Chel. etc. p. 285 ; id. Fauna Ind., Rept. p. 5 (skull fig. p. 2).

Cantor says that this species is numerous at Penang and on the coast of the Peninsula, but appears to be less so than *Crocodilus porosus.* There is a young specimen from Singapore in the British Museum.

Hab. India, Ceylon, Burma, Malay Peninsula and Archipelago.

Order SQUAMATA.

Suborder LACERTILIA.

Family GECKONIDÆ.

1. GYMNODACTYLUS AFFINIS, Stol.

Cyrtodactylus affinis, Stol. J. A. S. B. 1870, p. 167.
Gymnodactylus affinis, Boul. Cat. Liz. i. p. 42.

Stoliczka says, " The only specimen I caught between the bark of a large tree near the top of the Government bungalow on Penang hill ; " he subsequently found it in the collection he got from Penang and Province Wellesley (J. A. S. B. 1873, p. 113).
Hab. Malay Peninsula.

2. GYMNODACTYLUS PULCHELLUS, Gray.

Gymnodactylus pulchellus, Cantor, p. 25 ; Boul. Cat. Liz. i. p. 46.

Cantor says, " The species appears to be rather numerous on the hills at Penang, where the individuals obtained were captured in houses, at an elevation of 2200'." Stoliczka found it in the collection he got from Penang and Province Wellesley. There are specimens in the British Museum from Singapore. I obtained three specimens on Penang Hill at an elevation of 2200 ft. ; one was caught in an outbuilding, the other two in caves at night. Although so strikingly marked, they are very difficult to see in their natural surroundings, the colouring assimilates so to the irregular rocky walls of the caves. The largest specimen, ♂, was 220 mm. in total length (H.B. 113, tail 107).

Cantor's description of the life coloration is very good, but, as pointed out by Stoliczka[1], there are properly five dark bands across the neck and back (and not six). Cantor mentions these dark bands having sulphur or chrome-yellow margins, Stoliczka speaks of them as white-edged, and my specimens also had white margins. The upper surfaces of the limbs are uniform light yellowish brown, like the back ; and the under surfaces of my specimens were bluish buff.

As Cantor says, they bite fiercely when handled.

Hab. Bengal to Malay Peninsula.

3. GONATODES KENDALLI, Gray.

Gonatodes kendallii, Boul. Cat. Liz. i. p. 63.

This species, for some years only known from Borneo, was found in Perak by Mr. Wray, who sent a specimen from Larut (4200') to the British Museum, and in Singapore by Mr. Ridley, who sent ♂ and ♀ specimens to the British Museum. With his assistance I obtained this species at Singapore. It is to be found during the daytime in crevices under big rocks in the jungle on Bukit Timah, and it was only by burning paper in the crevices that we could get these active little Geckos to leave their retreats.

Hab. Malay Peninsula and Borneo.

4. GONATODES PENANGENSIS, n. sp.   (Plate XLIV. fig. 1.)

This species is very similar to *Gonatodes kendalli* in general appearance, but may be distinguished by the scaling of the lower side of the digits and by the presence of præanal pores in the male ; in this character connecting *G. kendalli* with the species, such as *G. ornatus*, which have præanal pores.

Description.—Habit very slender. Head oval ; snout broad and rounded, depressed, with the canthal ridges developed, longer than the distance between the eye and the ear-opening, nearly twice the diameter of the orbit. Eye large. Ear-opening vertically oval. Limbs long ; digits long and slender, compressed. The character

[1] J. A. S. B. 1873, p. 118.

of the scales on the lower surfaces of the digits of both hands and
feet at once separates this species from *G. kendalli*, in which
they are entirely covered with small transverse lamellæ, while in
this species, though the basal and terminal phalanges have transverse lamellæ, the intermediate one is covered with small irregular
scales. There is a large oval plate at the articulation of the basal
and proximal phalanges, as in *G. kendalli*. Upper surfaces
covered with minute granules, intermixed on the body with
irregularly arranged small tubercles, with slight keels. Rostral
large, quadrangular, not twice as broad as high, with median cleft.
Nostril between the rostral and several small scales. Nine to
thirteen upper and nine to eleven lower labials. Symphysial very
large, subtriangular. Two large chin-shields; two or three mental
scales following the symphysial are slightly enlarged. Abdominal
scales very small, juxtaposed, convex, keeled. Male with five to
seven prœanal pores, arranged in an obtusely angular or crescent-
shaped line. Tail cylindrical, slender, with small keeled scales and
some small pointed tubercles; but in none of the four specimens
examined are there the series of large spines which are to be seen
in some specimens of *G. kendalli*.

Colour (from life). Iris orange or yellow. Above yellowish
brown, mottled with dark brown, deepest (rich red-dark-brown)
on the shoulders. Five transverse yellow bands, two anterior
very bright, three posterior more or less indistinct in some
specimens; the tubercles on the two anterior yellow bands are of
a most brilliant golden colour. Below, head and throat bright
orange, remainder purplish-grey, shading to buff on chest and
extremities of limbs. Tail with alternate bands of light and dark
brown; in one specimen there are sixteen of these bands, and the
tubercles on the lighter bands are white. The sexes seem to be
coloured alike.

Size. The following dimensions are those of a male :—

	mm.
Total length	93
Head	$12\frac{1}{2}$
Width of head	$7\frac{1}{2}$
Body	$35\frac{1}{2}$
Fore limb	20
Hind limb	26
Tail	45

Females seem to be of about the same size.

Locality. I found these Geckos in March 1896, numerous in two
small caves in the rocks at the "Crag," Penang, at an elevation of
2200 feet: in which they were to be found running over the walls
both by day and night; at dusk they could also be found on rocks
in the open. They are very active.

Four specimens, three males and one female, are now in the
British Museum.

Hab. Malay Peninsula.

5. ÆLUROSCALABOTES FELINUS, Gthr.

Pentadactylus felinus, Günther, Rept. Brit. Ind. p. 117.
Ælurosaurus felinus, Boul. Cat. Liz. i. p. 73.
Æluroscalabotes felinus, Boul. op. cit. iii. p. 482.
The type specimen in the British Museum is from Singapore.
Hab. Malay Peninsula and Borneo.

6. HEMIDACTYLUS FRENATUS, D. & B.

Hemidactylus frenatus, Cantor, p. 23; Stol. J. A. S. B. 1870,
p. 104; Boul. Cat. Liz. i. p. 120.

Cantor says this species is "very numerous in valleys and hills ;
Malayan Peninsula, Pinang, and Singapore." Stoliczka says it
"occurs in Penang; I only obtained it on two occasions, both
times on the pillars of the verandah; it seems to have been
expelled from the interior apartments by the much stronger *Peripia
peronii"* (=*Gehyra mutilata*). I had not seen Stoliczka's paper
when I was at Penang, but arrived at the same conclusion about
the habits of this species; during a fortnight that I spent on
Penang Hill last March, I noticed that while *Gehyra mutilata*
swarmed in my room, the smaller *Hemidactylus frenatus* was only
to be found in the outer verandah. I also found this species in
Georgetown, Penang. The largest specimen I got was 106 mm.
in total length (H.B. 53 mm., tail 53 mm.).

Hab. Southern India, China, Indo-China, Malay Peninsula,
islands of Western Pacific and Indian Oceans and St. Helena.

7. HEMIDACTYLUS GLEADOVII, Murr.

Hemidactylus maculatus, part., Günther, Rept. Brit. Ind. p. 107.
Hemidactylus gleadovii, Boul. Cat. Liz. i. p. 129; id. Fauna Brit.
Ind., Rept. p. 86 (figured).

Günther mentions having seen specimens from Singapore.
Hab. India, Ceylon, Burma, South China, and Malay Peninsula.

8. HEMIDACTYLUS DEPRESSUS, Gray.

Hemidactylus depressus, Boul. Cat. Liz. i. p. 134.
There is a specimen from Singapore in the British Museum.
Hab. Ceylon, Malay Peninsula.

9. HEMIDACTYLUS LESCHENAULTI, D. & B.

Hemidactylus leschenaultii, Boul. Cat. Liz. i. p. 136.
There is a specimen in the British Museum from Penang from
Dr. Cantor.
Hab. India, Ceylon, Malay Peninsula.

10. HEMIDACTYLUS COCTÆI, D. & B.

Hemidactylus coctæi, Cantor, p. 23 ; Boul. Cat. Liz. i. p. 137.
Cantor observed two males in houses in the valley of Penang.
Hab. India, Malay Peninsula.

11. HEMIDACTYLUS PLATYURUS, Schn.

Nycteridium platyurus, Stol. J. A. S B. 1873, p. 113.
Nycteridium schneideri, Günther, Rep. Brit. Ind. p. 111.
Hemidactylus platyurus, Boul. Cat. Liz. i. p. 143.

Cantor observed this species in houses in the valley of Penang; Stoliczka found it in the collection he got from Penang and Province Wellesley.

Hab. India, Ceylon, South China, Indo-China, Malay Peninsula and Archipelago.

12. MIMETOZOON FLOWERI, Blgr.

Mimetozoon floweri, Boul. P. Z. S. 1896, p. 767, pl. xxxvi.

The specimen described by Mr. Boulenger I caught at dusk, running on the ground, in the garden of the "Crag" Hotel, Penang Hill, at an elevation of 2200', in March 1896.

Hab. Malay Peninsula.

13. GEHYRA MUTILATA, Wiegm.

Hemidactylus peronii, Cantor, p. 22.
Peripia peronii, Stoliczka, J. A. S. B. 1870, p. 163.
Gehyra mutilata, Boul. Cat. Liz. i. p. 148.

Cantor observed this species in houses in the valley of Penang; and Stoliczka says it is the most common House-Gecko all over the island of Penang, along the sea-coast as well as on the top of the hill, elevation 2500'. I found this Gecko swarming in houses wherever I stayed in Penang and Singapore (also in Deli, Sumatra), and, as Stoliczka says, from the sea-level to the top of the hill: it is to be found both inside and outside buildings, and I have also found it in gardens. It is very voracious, and will attempt to seize any insect; I have more than once seen a *Gehyra* attack a full-sized *Hierodula vitrea* and repulsed. It shows great ingenuity both in escaping capture and in obtaining its food. It frequents lamps especially at night, to catch the insects attracted by the light. Whenever these Geckos are about you hear their cheerful noise, and also at intervals during the day when they are out of sight in holes or under the roof. Cantor (p. 20), in describing *Gecko monarchus*, says its cry resembles the monosyllable 'tok,' repeated 6 or 8 times with increased celerity; I have not heard the cry of *G. monarchus*, but the above description well suits that of *Gehyra mutilata*.

These Geckos throw off their tails on the slightest provocation. There was one living in the Officers' Mess at Penang, in which the reproduced tail had grown bifid laterally.

The young are very different in appearance to the adults, on account of the slenderer body and tail and the coloration. Stoliczka (J. A. S. B. 1870, p. 163) says "the young lizard is brown, with numerous rather large round pale spots all over the body;" but I have found them pale olive-brown with distinct dark brown spots above, and immaculate buff beneath. They seem to vary greatly. The spots disappear as they grow larger. The adults

have the power of changing their colour to some extent (as Cantor
remarks of *Gecko monarchus*); they are generally buff or ash-
coloured by day and almost white by night. From measuring a
large series of specimens, I should say the total length of a full-
grown average *G. mutilata* is 120 mm., of which the head and body
and the tail are each about half.

Hab. Mascarene Islands, Seychelles, Ceylon, Burma, Malay
Peninsula and Archipelago, New Guinea; Western Mexico.

14. LEPIDODACTYLUS CEYLONENSIS, Blgr.

Lepidodactylus ceylonensis, Boul. Cat. Liz. i. p. 164, pl. xiii. fig. 3.

This species appears not to have been previously recorded from
the Straits Settlements. I caught one female in Headquarter
House, Singapore. Total length 65 mm. (H.B. 36, tail 29).

Colour. Above dark brown, spotted with brick-red and black.
Black lateral line from snout to shoulder passing through eye.
Light yellow spots on lips and behind eyes. Upper surface of
tail red with brown marks. The underpart of the body was
purplish-brown, of the tail rusty-brown and yellow, with minute
black spots.

Hab. Ceylon, Burma, Malay Peninsula, Borneo, Java, and
Engano.

15. LEPIDODACTYLUS LUGUBRIS, D. & B.

Platydactylus lugubris, Cantor, p. 16.
Peripia cantoris, Günther, Rept. Brit. Ind. p. 110.
Lepidodactylus lugubris, Boul. Cat. Liz. i. p. 165.

Cantor says a single male was captured in his house in the
valley of Penang.

Hab. Malay Peninsula and Archipelago, New Guinea, and
Polynesia.

16. GECKO VERTICILLATUS, Lawr.

Platydactylus gecko, Cantor, p. 17.
Gecko guttatus, Günther and Stol.
Gecko verticillatus, Boul. Cat. Liz. i. p. 183.

There is a specimen in the British Museum from the Malay
Peninsula through Dr. Cantor. Stoliczka found it in the collection
he got from Penang and Province Wellesley. Müller mentions
Gecko guttatus in the Bâle Museum from Singapore; and
Dr. Blanford (P. Z. S. 1881, p. 215) mentions it in the collection he
got from Dr. Dennys from Singapore and neighbouring localities.

Hab. N.E. India, Burma, Southern China, Anam, Siam, Malay
Peninsula and Archipelago.

17. GECKO STENTOR, Cant.

Platydactylus stentor, Cantor, p. 18.
Gecko stentor, Günther, Rept. Brit. Ind. p. 102, pl. xi. fig. A;
Boul. Cat. Liz. i. p. 184.

Cantor obtained the type specimen "from the villa on the

Pentland Hills, Penang." Stoliczka (J. A. S. B. 1873, p. 113)
found it in the collection he got from Penang and Province
Wellesley, and also (J. A. S. B. 1870, pp. 161, 162) mentions a
specimen of *Gecko smithii* in the Fort Pit Museum said to be from
Penang.

Hab. Burma, Andamans, Malay Peninsula and Archipelago.

18. GECKO MONARCHUS, D. & B.

Platydactylus monarchus, Cantor, p. 19.
Gecko monarchus, Boul. Cat. Liz. i. p. 187.

Cantor says, "In the valleys and on the hills of Penang it is
very numerous, swarming at night in rooms, on the walls, and
under the ceiling;" he also mentions the Malay Peninsula and
Singapore as localities. Müller records this species from Singapore
in the Bâle Museum. There are specimens in the British Museum
from Penang and Singapore. I did not see this Gecko in Penang,
but I found it common about the aviary in the Botanical Gardens,
Singapore.

Hab. Ceylon, Malay Peninsula and Archipelago.

19. PTYCHOZOON HOMALOCEPHALUM, Crev.

Ptychozoon homalocephalum, Cantor, p. 20; Stol. J. A. S. B.
1870, p. 159.
Ptychozoon homalocephalum (part.), Boul. Cat. Liz. i. p. 190.

Cantor mentions two specimens captured in a villa on Penang
Hill. Stoliczka says it is not uncommon in Penang, but from
what I heard from inhabitants it must at any rate be rarely seen.

Hab. Burma, Malay Peninsula and Archipelago.

20. PTYCHOZOON HORSFIELDI, Gray.

Ptychozoon homalocephalum (part.), Boul. Cat. Liz. i. p. 190.
Ptychozoon horsfieldi, F. Müller, Ver. nat. Ges. Basel, 1892,
p. 209, pl. iv.

The type specimen, a female, is from Singapore from General
Hardwicke's collection. I caught a male on a wooden post in the
Experimental Gardens, Penang Hill, 1900 ft. Total length 155 mm.
(H.B. 80, tail 75). It tried to bite fiercely when handled. Both
specimens are now in the British Museum.

Hab. Malay Peninsula and Archipelago.

Family AGAMIDÆ.

21. DRACO VOLANS, L.

Draco volans, Boul. Cat. Liz. i. p. 256.

Cantor gives Malayan Peninsula and Penang as localities for
this species. Stoliczka says it "appears to be more common in
the jungles of the Wellesley Province and near Malacca, than it is
on Penang itself." Mr. Mitchell gave me two specimens caught
at Kulim, Kedah. In February, March, and April of this year
I found this species very numerous about Tanglin, Singapore;

males were more plentiful than females. Of over twenty specimens examined, the largest male was 200 mm. in total length (H.B. 77, tail 123), and 86 mm. in extent across its extended " wings "; and the largest female was 193 mm. in total length (H.B. 75, tail 118). Some of the females contained four rich-yellow-coloured leathery-skinned eggs about 5 by 4 mm.

Mr. Ridley found this species at the Dindings.

Life-coloration.—In Cantor's description he does not mention the differences between the male and female; in my specimens I found these both noticeable and constant.

Male. Front part of upper surface of head sea-green, with a black spot between the eyes. General colour of upper surfaces light bronze-brown, mottled all over with spots and patches of rich red-brown, dark brown, and black; in some lights fugitive metallic green shades are seen. Some of the markings are more definite than others : these are a median black spot on the nape of the neck, a cluster of black spots in front of the shoulders, two broken transverse black bands across the body, and a pair of black spots in front of hips.

Under surfaces of head, body, and limbs are brownish-buff minutely spotted with dark brown, and metallic green shades are frequent and vivid. The gular pouch is brilliant yellow.

Tail bronze-brown with rings of dark brown.

Wing-membrane—of the upper surface the portion nearest the body is of the same light bronze-brown as the back, mottled with dark brown, beyond this the ground-colour is orange-red, and the markings get larger and darker, till towards the margin they coalesce and the light ground-colour cannot be seen. Round the margin of the parachute is a narrow border of light brown speckled with black. The under surface varies from pale cobalt to bright blue, with pink patches and large bars and dots of black.

Female. Differing from the male as follows :—

(i.) Front part of upper surface of head very dark brown or grey (black spot as in male).

(ii.) The gular pouch is blue or green, minutely speckled with black.

(iii.) Where the ground-colour of the upper surface of the wing-membrane is orange-red in the male, it is rich yellow in the female.

(iv.) The under surface of the wing-membrane is greenish-yellow, there are no pink patches, and the black bars and spots are larger.

These Lizards when at rest on the trunk of a tree, usually in a vertical position, are almost invisible, owing to their dark mottled-brown colour, but when darting through the air overhead they resemble a flashing blue gem, owing to the bright colours of the underneath of the " wings." They are very active and nimble, spreading their parachute as they leap from any point, and alighting gently on all fours closing it as they touch the ground. They can apparently direct their flight exactly. I have seen one slide through the air (with its wings quite steady) for a distance of about 20 yards, and then settle on the trunk of a tree.

Hab. Malay Peninsula and Archipelago.

22. DRACO MACULATUS, Gray.

Draco maculatus, Cantor, p. 39; Boul. Cat. Liz. i. p. 262.

Cantor obtained four specimens from the hills of Penang.

Hab. Assam, Burma, Camboja, Malay Peninsula.

23. DRACO FIMBRIATUS, Kuhl.

Draco fimbriatus, Stol. J. A. S. B. 1873, p. 119; Boul. Cat. Liz. i. p. 265.

There is a specimen from Singapore in the British Museum, and Stoliczka found a specimen in the collection he got from Penang and Province Wellesley.

Hab. Malay Peninsula and Archipelago.

24. DRACO QUINQUEFASCIATUS, Gray.

Draco quinquefasciatus, Stol. J. A. S. B. 1873, p. 118; Boul. Cat. Liz. i. p. 269, pl. xx. fig. 8.

The type specimen in the British Museum is from Penang, from Gen. Hardwicke's collection, and Stoliczka obtained one specimen in the collection he got from Penang and Province Wellesley.

Hab. Malay Peninsula and Borneo.

25. DRACO MELANOPOGON, Blgr.

Draco melanopogon, Boul. Cat. Liz. iii. p. 492.

The types are in the British Museum; they are from Malacca, presented by Mr. Hervey.

Hab. Malay Peninsula, Borneo, and Natunas.

26. APHANIOTIS FUSCA, Ptrs.

Aphaniotis fusca, Boul. Cat. Liz. i. p. 274.

There are two specimens in the British Museum from Malacca, presented by Mr. Hervey.

Hab. Malay Peninsula, Borneo, and Natunas.

27. GONYOCEPHALUS HERVEYI, Blgr.

Gonyocephalus herveyi, Boul. Cat. Liz. iii. p. 493.

The type specimen is in the British Museum from Malacca, presented by Mr. Hervey.

Hab. Malay Peninsula and Natunas.

28. GONYOCEPHALUS BORNEENSIS, Schl.

Gonyocephalus borneensis, Boul. Cat. Liz. i. p. 288.

There are four specimens in the British Museum from Malacca, presented by Mr. Hervey.

Hab. Malay Peninsula and Borneo.

29. GONYOCEPHALUS GRANDIS, Gray.

Diplophyrus grandis, Cantor, p. 34, pl. xx.
Gonyocephalus grandis, Boul. Cat. Liz. i. p. 298.

Cantor obtained one specimen from the hills of Penang, at an elevation of 2000 feet.

Hab. Burma, Malay Peninsula, Sumatra, and Borneo.

30. ACANTHOSAURA ARMATA, Gray.

Lophyrus armatus, Cantor, p. 32.
Acanthosaura armata, Boul. Cat. Liz. i. p. 301, pl. xxii. fig. 1.

Cantor says that "two individuals were obtained from spice plantations in the valley" at Penang, and there are specimens in the British Museum from Singapore from Gen. Hardwicke's collection.

Hab. Tenasserim, Siam, Cochinchina, and Malay Peninsula.

31. CALOTES CRISTATELLUS, Kuhl.

Bronchocela cristatella, Cantor, p. 30.
Calotes cristatellus, Boul. Cat. Liz. i. p. 316.

Cantor says, "This species is very numerous in the Malayan countries both in the valleys and on the hills, Malayan Peninsula, Pinang, Singapore." Stoliczka obtained specimens from Penang and Province Wellesley, and from Singapore (*Bronchocela moluccana*). Dr. Blanford mentions it in the collection he got from Dr. Dennys from Singapore. I only obtained one specimen in Penang, but at Tanglin, Singapore, found this species very numerous; the largest male was 481 mm. in total length (H.B. 113, tail 368), the females seem to grow to nearly the same size. In one specimen caught at Tanglin, the tail at 113 mm. from the anus bifurcated, one branch being 109 mm. long from fork to tip, the other 197 mm.

As Cantor remarks about this Lizard, "the most striking feature is the great power of suddenly changing its colours." Both this species and *Calotes versicolor* are commonly called Chameleons by the English in the Straits Settlements. Among the Klings in Singapore there is a belief that these Lizards have twelve different colours, which they change during the day, a colour for every hour.

The colours I have noticed of this species are:—

    (i.) Very light yellowish-green.
    (ii.) Bright grass-green.
    (iii.) Bright green as above with large dark-brown patches.
    (iv.) Dark green, almost black.
    (v.) Dark brown, almost black.
    (vi.) Dull grey-brown.

The brighter green colours are generally uniform; but the other colours are on the neck, back, and sides irregularly spotted or reticulated with darker; or else there are dark bands longitudinal on the neck and transverse on the body; while the limbs and tail are usually marked with transverse dark brown irregular bands.

In April the lips, cheeks, and throat of the males were very beautiful with golden, red, and crimson shades on the scales.

Both this species and *C. versicolor* seem of similar habits, liking bright sunshine and frequenting gardens and cultivated open land with small bushes, darting about the grass and climbing the branches with the utmost agility. When caught they try to defend themselves by biting fiercely.

Hab. Tenasserim, Malay Peninsula and Archipelago.

32. CALOTES VERSICOLOR, Daud.

Calotes versicolor, Boul. Cat. Liz. i. p. 321, & Fauna Brit. Ind., Rept. p. 135, fig. p. 136.

Neither Cantor nor Stoliczka seem to have observed this species in the Malay countries. F. Müller records it from Penang in the Bâle Museum, and Blanford mentions it in the collection he got from Dr. Dennys from Singapore and neighbouring localities.

I found this species fairly common about the Sepoy Lines, Penang; a female caught in March contained seven white leathery-skinned eggs, and one caught in April contained eight. In the newly-cleared country around Kulim, Kedah, there were large numbers of *Calotes*; when the jungle has been cut down, stumps of the larger forest trees are left standing here and there, several yards high out of the ground; on a bright sunshiny day, a *Calotes* was to be seen on the summit of nearly every one of these stumps, apparently enjoying the warmth and waiting for passing insects. The only specimen I obtained here was of this species.

Hab. Afghanistan, Beloochistan, India, Ceylon, Burma, Southern China, Siam, and the Malay Peninsula.

33. LIOLEPIS BELLII, Gray.

Liolepis bellii, Cantor, p. 41; Boul. Cat. Liz. i. p. 403.

There are specimens in the British Museum from Penang, from Cantor and Capt. Stafford. Cantor says, "This species appears to be numerous, but local. Twelve were at one time obtained from a spice plantation in Province Wellesley."

Hab. Southern India, Burma, Southern China, Siam, and Malay Peninsula.

Family VARANIDÆ.

34. VARANUS FLAVESCENS, Gray.

Varanus flavescens, Cantor, p. 28; Günther, Rept. Brit. Ind. p. 65, pl. ix. fig. A; Boul. Cat. Liz. ii. p. 309.

Cantor obtained a single specimen at Penang.

Hab. Northern India, Burma, Malay Peninsula.

35. VARANUS NEBULOSUS, Gray.

Varanus nebulosus, Cantor, p. 27; Günther, Rept. Brit. Ind. p. 66, pl. ix. fig. D; Boul. Cat. Liz. ii. p. 311.

Cantor obtained one specimen in the hills of Penang; there are in the British Museum three specimens from Malacca from Mr. Hervey.

Hab. Bengal, Burma, Siam, and Malay Peninsula.

36. VARANUS RUDICOLLIS, Gray.

Varanus rudicollis, Boul. Cat. Liz. ii. p. 313.

There is a specimen in the British Museum from Malacca from Mr. Hervey.

Hab. Malay Peninsula, Borneo, Philippines.

37. VARANUS SALVATOR, Laur.

Hydrosaurus salvator, Günther, Rept. Brit. Ind. p. 67, pl. ix. fig. E.

Varanus salvator, Cantor, p. 29; Boul. Cat. Liz. ii. p. 314; Boul. Fauna Brit. Ind., Rept. p. 166 (head fig. p. 162).

Cantor says, " This species is very numerous both in hilly and marshy localities; Malayan Peninsula and Pinang." Stoliczka found it in the collection he got from Penang and Province Wellesley. Dr. Blanford found it in the collection he got from Dr. Dennys from Singapore. I saw many of these Lizards on the Kedah river in April 1895, and obtained one from Blakan Mati, Singapore, in January 1896. The English in India and the Straits Settlements call them "Iguana," and the Malays "Beyawh." The Chinese prize them highly for the supposed medicinal properties of the heart, liver, etc. These Lizards are generally infested with ticks, much resembling one of their scales in size and colour. A great part of their food seems to consist of the small crabs which abound on the mud of the mangrove swamps. In life they are very handsomely marked—black and bright yellow. The largest specimen obtained I shot in the Gunong Gajah tributary of the Kedah river. It was a male—Total length 2362 mm.; head and body 1041; tail 1321; girth behind forearms 470; girth round stomach 584. It is now mounted in the Reptile Gallery of the British Museum.

Hab. Nepaul, Ceylon, China, Siam, Tenasserim, Malay Peninsula and Archipelago, Cape York.

Family SCINCIDÆ.

38. MABUIA NOVEMCARINATA, And.

Mabuia novemcarinata, Boul. Cat. Liz. iii. p. 179.

This species was discovered by Dr. Anderson in Burma. It can now be added to the list of Malayan reptiles, as I caught a specimen near "the Crag," Penang Hill, elevation 2200 ft., in March

1896. The colour of the upper parts was bronze, a black band along each side, and the belly pale green. Total length 205 mm. (H.B. 92, tail 113).

Hab. Burma, Malay Peninsula.

39. MABUIA MULTIFASCIATA, Kuhl.

Euprepes rufescens, Cantor, p. 46.
Tiliqua carinata, part., Stol. J. A. S. B. 1870, p. 169.
Mabouia multifasciata, Boul. Cat. Liz. iii. p. 186.

Cantor says it is " exceedingly numerous in the hills and valleys of the Malayan countries. Peninsula, Pinang, and Singapore." Stoliczka found it common at Penang and on the coast of Province Wellesley. This species is very common about George-town, Penang, especially when the sun comes out after heavy rain, large numbers are to be seen in the grass and on the stone edges of the surface drains, enjoying the warmth and showing off their brilliant metallic colours. I obtained several specimens at Singapore, but did not see it in the same abundance as at Penang. They vary a good deal in colour, but the most usual variety has the upper parts uniform olive-brown or bronze, and the lower parts pale greenish-yellow, with on either side a broad red stripe starting from above and behind the tympanum, and continuing either halfway down the body or to the hip ; this stripe is highly iridescent, and changes to gold, orange, crimson, and green as the light plays on the living animal. The largest specimen obtained (from Singapore) was 314 mm. in total length (H.B. 111, tail 203).

Hab. Eastern Himalayas, Burma, Siam, Malay Peninsula and Archipelago.

40. LYGOSOMA ANOMALOPUS, Blgr.

Lygosoma anomalopus, Boul. P. Z. S. 1890, p. 84, pl. xi. fig. 4.

There are two specimens in the British Museum from Dr. J. G. Fischer from Penang.

Hab. Malay Peninsula and Sumatra.

41. LYGOSOMA OLIVACEUM, Gray.

Euprepis ernestii, Cantor, p. 47.
Lygosoma olivaceum, Boul. Cat. Liz. iii. p. 251.

Cantor mentions this species from the Peninsula and Penang. Stoliczka found it in the collection he got from Penang and Province Wellesley.

The young of this species is very brightly coloured, as mentioned by Cantor (p. 48) and by Stoliczka (J. A. S. B. 1873, p. 118). Although the general scheme of marking is the same, individuals apparently vary, so, to compare with the above accounts, I give the colours of a specimen caught by Mr. Ridley in a coco-nut tree at Galang, Singapore, last April. The length of head and

body was 32 mm., and the tail (of which the tip was broken)
36 mm. The upper surface of head was light brown, the scales
being edged with black lines. A black line through eye. Lips
and chin immaculate black. The back, sides, and upper surfaces
of the limbs were black, with wavy, irregular but well-defined
transverse lines, pale greenish-white anteriorly, gradually getting
yellowish further back till those across the base of the tail were
yellowish-red. The lower surfaces of the body and limbs were
greenish-white. The lower surfaces of the toes and palms of feet
were brown. The tail was bright red, paler beneath.
Hab. Tenasserim, Nicobars, Malay Peninsula and Archipelago.

42. LYGOSOMA SINGAPORENSE, Stdr.

Eumeces (Mabouya) singaporensis, Steindachn. Sitzb. Ak. Wien,
lxii. i. 1870, p. 341, pl. iv. fig. 2.
Lygosoma singaporense, Boul. Cat. Liz. iii. p. 297.

Steindachner described this species from a specimen from Singapore.
Hab. Malay Peninsula.

43. LYGOSOMA JERDONIANUM, Stol.

Mabouya jerdoniana, Stol. J. A. S. B. 1870, p. 172.
Lygosoma jerdonianum, Boul. Cat. Liz. iii. p. 300.

Stoliczka caught the type specimen and saw others on the little
island of Pulo Tikus, Penang, and mentions having observed a
very similar, or the same species on one of the small islands near
Singapore.
Hab. Malay Peninsula.

44. LYGOSOMA BOWRINGII, Gthr.

Eumeces bowringii, Günther, Rept. Brit. India, p. 91.
Euprepes (Riopa) punctatostriatus, Peters, Mon. Berl. Ac. 1871,
p. 31.
Lygosoma bowringii, Boul. Cat. Liz. iii. p. 308, pl. xxiii. fig. 3.

Peters records a specimen from Singapore.
Hab. Burma, Hongkong, and Malay Peninsula.

45. LYGOSOMA ALBOPUNCTATUM, Gray.

Eumeces punctatus, Cantor, p. 45.
Riopa albopunctata, Stoliczka.
Lygosoma albopunctatum, Boul. Cat. Liz. iii. p. 309.

Cantor says, it " is numerous in the Malayan countries, both on
hills and in valleys. Peninsula, Pinang, and Singapore." Stoliczka
found it in the collection he got from Penang and Province
Wellesley.
Hab. India, Assam, Burma, and Malay Peninsula.

46. LYGOSOMA CHALCIDES, L.

Lygosoma chalcides, Cantor, p. 49 ; Boul. Cat. Liz. iii. p. 340.

Cantor obtained two specimens on Penang Hill, and mentions a third from Singapore in the Museum of the Asiatic Society.

Hab. Southern China, Siam, Malay Peninsula, and Java.

NOTE.—In the collection sent by Dr. Dennys from Singapore, and described by Dr. Blanford (P. Z. S. 1881, p. 215), occurs *Eumeces chinensis* ; but as these specimens were from the Raffles Museum, and their locality not known, it probably was not caught in the Malayan countries, but brought from China.

Suborder OPHIDIA.

Family TYPHLOPIDÆ.

1. TYPHLOPS LINEATUS, Boie.

Pilidion lineatum, Cantor, p. 50.
Typhlina lineata, Günth. Rep. Brit. Ind. p. 171, pl. xvi. fig. B.
Typhlops lineatus, Boul. Cat. Snakes, i. p. 15.

I obtained one specimen on Penang Hill, 2200 feet. Cantor mentions it from the hills of Penang, and there are specimens in the British Museum from Singapore and Malacca.

Hab. Malay Peninsula and Archipelago.

2. TYPHLOPS BRAMINUS, Daud.

Typhlops braminus, Cantor, p. 52 ; Günth. Rept. Brit. Ind. p. 175, pl. xvi. fig. 1 ; Stol. J. A. S. B. 1873, p. 114 ; Boul. Cat. Snakes, i. p. 16.

I obtained one specimen on Penang Hill, 2200 ft. Cantor says it is " numerous in hills and valleys, Pinang, Singapore, Malayan Peninsula." Stoliczka found it in the collection he got from Penang and the Province Wellesley. Mr. Ridley has found it at Singapore.

Hab. South Asia ; islands of the Indian Ocean ; Africa south of the Equator.

3. TYPHLOPS BOTHRIORHYNCHUS, Gthr.

Typhlops bothriorhynchus, Günth. Rept. Brit. Ind. p. 174, pl. xvi. fig. G ; Boul. Cat. Snakes, i. p. 23.

The type specimen is in the British Museum from Penang, from Dr. Cantor.

Hab. Northern India (North-West Provinces and Assam) and Malay Peninsula.

4. TYPHLOPS NIGROALBUS, D. & B.

Typhlops nigroalbus, Cantor, p. 51 ; Günth. Rept. Brit. Ind. p. 172, pl. xvi. fig. F ; Stol. J. A. S. B. 1873, p. 114 ; Boul. Cat. Snakes, i. p. 24.

I obtained one specimen on Penang Hill, 2200 ft. Length

136 mm. Colour, upper parts black, highly iridescent, lower parts pinky-grey.

Cantor found two specimens on Penang Hill ; Stoliczka found it in the collection he got from Penang and Province Wellesley ; and there are specimens in the British Museum from Perak and Singapore.

Hab. Malay Peninsula and Sumatra.

Family BOIDÆ.

5. PYTHON RETICULATUS, Schn.

Python reticulatus, Cantor, p. 55 ; Stol. J. A. S. B. 1873, p. 115 ; Boul. Cat. Snakes, i. p. 85.

Cantor says this species is " very numerous in the Malayan hills and valleys " ; Stoliczka found it in the collection he got from Penang and Province Wellesley, and there are specimens in the British Museum from Penang and Singapore. I have seen specimens recently caught in Penang, on the Krean river (Prov. Wellesley), and near Taiping, Perak.

Hab. Burma, Indo-China, Malay Peninsula and Archipelago.

6. PYTHON MOLURUS, L.

Python molurus, Stol. J. A. S. B. 1870, p. 205 ; Boul. Cat. Snakes, i. p. 87.

Stoliczka mentions having "seen several specimens obtained in the Wellesley province."

Hab. India, Ceylon, Southern China, Malay Peninsula, and Java.

7. PYTHON CURTUS, Schl.

Python curtus, Boul. Cat. Snakes, i. p. 89, and P. Z. S. 1889, pl. xlv.

There are specimens in the British Museum from Malacca and Singapore.

Hab. Malay Peninsula, Sumatra, and Borneo.

Family ILYSIIDÆ.

8. CYLINDROPHIS RUFUS, Lawr.

Cylindrophis rufus, Cantor, p. 53 ; Stol. J. A. S. B. 1873, p.114 ; Boul. Cat. Snakes, i. p. 135.

I obtained one specimen at Tanglin, Singapore. Length 483 mm. Scales in 21 rows. Colour, uniform black above, belly black with transverse white bands, orange collar-mark on neck, bright vermilion mark on tail. A second specimen obtained in Singapore was 546 mm. in length. Cantor mentions one specimen from Singapore. Stoliczka found it in the collection he got from Penang and Province Wellesley, and there are specimens in the British Museum from Penang and Singapore.

Hab. Burma and Cochinchina to the Malay Peninsula and Archipelago.

9. CYLINDROPHIS LINEATUS, Blanf.

Cylindrophis lineatus, Blanford, P. Z. S. 1881, p. 217, pl. xx.;
Boul. Cat. Snakes, i. p. 137.

The type specimen, from Singapore, belonging to the Raffles
Museum, was described by Mr. Blanford in 1881.

Hab. Malay Peninsula.

Family XENOPELTIDÆ.

10. XENOPELTIS UNICOLOR, Reinw.

Xenopeltis unicolor, Cantor, p. 54; Peters, Monatsb. Ak. der
Wiss. zu Berlin, 1859, p. 269; Boul. Cat. Snakes, i. p. 168 (skull
figured).

Cantor mentions this species from Penang Hill, Province
Wellesley, and Singapore; there is a specimen in the British
Museum from Singapore from Dr. Dennys. Peters mentions a
specimen from Princess Hill, Singapore. Of two specimens
observed by me in Singapore, the first, from Tanjong Katong, had
ventrals 188, subcaudals 34, and was 4·14 mm. in length; the second,
from Tanglin, had ventrals 175, subcaudals 34, and was 875 mm.
in length.

Hab. Southern India, Burma, Indo-China, Malay Peninsula and
Archipelago.

Family COLUBRIDÆ.

Series Aglypha.

Subfamily ACROCHORDINÆ.

11. ACROCHORDUS JAVANICUS, Hornst.

Acrochordus javanicus, Cantor, p. 58; Boul. Cat. Snakes, i.
p. 173.

Cantor mentions this species from Penang Hill and Singapore.
Hab. Malay Peninsula, Java, and New Guinea.

12. CHERSYDRUS GRANULATUS, Schn.

Acrochordus granulatus, Cantor, p. 59.
Chersydrus granulatus, Boul. Cat. Snakes, i. p. 174.

There is a specimen in the British Museum from Penang from
Dr. Cantor, and one from Singapore from Gen. Hardwicke.
Hab. From Southern India and Cochinchina to New Guinea.

13. XENODERMUS JAVANICUS, Reinh.

Xenodermus javanicus, Boul. Cat. Snakes, i. p. 175.

"Our collection contains also a specimen from Penang."—
F. Müller, Verh. nat. Ges. Basel, 1887, p. 268.

Hab. Malay Peninsula, Sumatra, Java.

Subfamily COLUBRIN.E.

14. POLYODONTOPHIS GEMINATUS, Boie.

Herpetodryas prionotus, Cantor, P. Z. S. 1839, p. 52.
Ablabes melanocephalus, Günther, R. B. I. p. 229.
Polyodontophis geminatus, Boul. Cat. Snakes, i. p. 185.

Cantor mentions a specimen from Malacca. Stoliczka[1] found one example in the Botanical Gardens at Singapore, and there are specimens in the British Museum from Singapore from General Hardwicke and Dr. Dennys.

Hab. Siam, Malay Peninsula, Sumatra, Java, and Borneo.

15. POLYODONTOPHIS SAGITTARIUS, Cant.

Calamaria sagittaria, Cantor, p. 64.
Polyodontophis sagittarius, Boul. Cat. Snakes, i. p. 187.

Cantor mentions one specimen from the Malay Peninsula.
Hab. West Himalayas, Bengal, Assam, Malay Peninsula.

16. XENOCHROPHIS CERASOGASTER, Cant.

Tropidonotus cerasogaster, Cantor, p. 92.
Xenochrophis cerasogaster, Boul. Cat. Snakes, i. p. 191.

Cantor mentions one specimen from the Province Wellesley.
Hab. Bengal, Assam, Khasi Hills, and Malay Peninsula.

17. TROPIDONOTUS TRIANGULIGERUS, Boie.

Tropidonotus umbratus, part., Cantor, p. 89.
Tropidonotus trianguligerus, Boul. Cat. Snakes, i. p. 224.

Stoliczka found this Snake in the collection he got from Penang and Province Wellesley, and there are specimens in the British Museum from Penang and Singapore.

Hab. Southern Burma, Malay Peninsula, Sumatra, Borneo, Java, and Ternate.

18. TROPIDONOTUS PISCATOR, Schn.

Tropidonotus umbratus, part., Cantor, p. 89.
Tropidonotus piscator, Boul. Cat. Snakes, i. p. 230.

Stoliczka found this Snake (*T. quincunctiatus*) in the collection he got from Penang and Province Wellesley.

Var. A. There is a specimen in the British Museum from Singapore.

Var. B. There is a specimen in the British Museum from Penang from Dr. Cantor. I obtained one specimen from the Racecourse, Penang. Ventrals 125; subcaudals 77.

Hab. India, Burma, Southern China, Indo-China, Malay Peninsula and Archipelago.

[1] J. A. S. B. xxxix. part ii. 1870, p. 183.

19. TROPIDONOTUS STOLATUS, L.

Tropidonotus stolatus, Cantor, p. 90 ; Boul. Cat. Snakes, i. p. 253.

Cantor mentions this species from the Malay Peninsula ; and there is a specimen in the British Museum from Singapore from Dr. Dennys.

Hab. India, Ceylon, Burma, China, Malay Peninsula, and Philippine Islands.

20. TROPIDONOTUS VITTATUS, L.

Tropidonotus vittatus, Boul. Cat. Snakes, i. p. 255.

Stoliczka mentions *T. vittatus* (Günther's ' Colubrine Snakes ') as occurring in the collection he got from Penang and Province Wellesley.

Hab. Malay Peninsula, Java, Celebes.

21. TROPIDONOTUS SUBMINIATUS, Schl.

Tropidonotus subminiatus, Boul. Cat. Snakes, i. p. 256.

This Snake is said to be found in the Malay Peninsula, and as it is recorded from Tenasserim and Java it seems probable.

Hab. From the Eastern Himalayas, Assam, Burma, and Southern China to the Malay Peninsula and Archipelago.

22. TROPIDONOTUS CHRYSARGUS, Schl.

Tropidonotus junceus, Cantor, p. 93.
Tropidonotus chrysargus, Boul. Cat. Snakes, i. p. 258.

There are specimens in the British Museum from Penang Hill from Dr. Cantor, and from Perak (hills over 3000 feet) from Mr. Wray.

Hab. From the Eastern Himalayas, Assam, Burma, and Southern China to the Malay Peninsula and Archipelago.

23. TROPIDONOTUS MACULATUS, Edel.

Tropidonotus maculatus, Boul. Cat. Snakes, i. p. 260.

There is a specimen in the British Museum from Malacca from Mr. Hervey.

Hab. Malay Peninsula, Sumatra, and Borneo.

24. MACROPISTHODON FLAVICEPS, D. & B.

Macropisthodon flaviceps, Boul. Cat. Snakes, i. p. 266.

There are two specimens in the British Museum from Perak from Mr. Wray.

Hab. Malay Peninsula, Sumatra, and Borneo.

25. MACROPISTHODON RHODOMELAS, Boie.

Macropisthodon rhodomelas, Boul. Cat. Snakes, i. p. 266.

There are several specimens in the British Museum from Singapore, where this Snake is very common. Between January and April 1896 I came across about fifteen specimens around

Tanglin, Singapore, mostly found in short grass and among low bushes; the largest was 609 mm. in length.

Hab. Malay Peninsula, Sumatra, Borueo, and Java.

26. HELICOPS SCHISTOSUS, Daud.

Tropidonotus schistosus, Cantor, p. 91.
Helicops schistosus, Boul. Cat. Suakes, i. p. 274.

Cantor mentions this species from the Malay Peninsula.

Hab. Southern India, Ceylon, Bengal, Burma, Yunnan, and Malay Peninsula.

27. LYCODON AULICUS, L.

Lycodon aulicus, part., Cantor, p. 68; Blanford, P. Z. S. 1881, p. 215; Boul. Cat. Snakes, i. p. 352.

Cantor mentions this species from Penang and the Malay Peninsula; Stoliczka found it in the collection he got from Penang and Province Wellesley; and Blanford mentions it from Singapore.

Hab. India, Ceylon, Himalayas, Burma, Siam, Cochinchina, Malay Peninsula and Archipelago.

28. LYCODON EFFRENIS, Cant.

Lycodon effrenis, Cantor, p. 70, pl. xl. fig. 2; Boul. Cat. Snakes, i. p. 356.

Cantor obtained one specimen from Penang Hill.

Hab. Malay Peninsula, Sumatra, and Borneo.

29. LYCODON SUBCINCTUS, Boie.

Lycodon platurinus, Cantor, p. 96.
Lycodon subcinctus, Boul. Cat. Snakes, i. p. 359.

Cantor mentions one specimen from Penang Hill, and there are two specimens in the British Museum from Singapore; I obtained one specimen at Singapore. Ventrals 202; subcaudals 85; length 635 mm.

Hab. Malay Peninsula, Sumatra, Borneo, Java, Philippines.

30. DRYOCALAMUS SUBANNULATUS, D. & B.

Nymphophidium subannulatum, Blanford, P. Z. S. 1881, p. 219.
Dryocalamus subannulatus, Boul. Cat. Snakes, i. p. 371.

There is a specimen in the British Museum from Singapore from Mr. Ridley; and Blanford mentions a specimen from Singapore belonging to the Raffles Museum; I obtained one from Province Wellesley.

Hab. Malay Peninsula and Sumatra.

31. ZAOCYS CARINATUS, Gthr.

Zaocys carinatus, Boul. Cat. Snakes, i. p. 377, pl. xxvii. fig. 1.

There are specimens in the British Museum from Perak and Singapore.

Hab. Malay Peninsula, Sumatra, and Borneo.

57*

32. ZAMENIS KORROS, Schl.

Coluber korros, Cantor, p. 74.
Zamenis korros, Boul. Cat. Snakes, i. p. 384.

1 obtained one specimen near Taiping, Perak. Cantor records
it from Penang, Singapore, and the Peninsula; Stoliczka found it
in the collection he got from Penang and Province Wellesley; and
there are specimens in the British Museum from Penang and
Singapore.

Hab. Sikhim Himalayas, Assam, Burma, Western Yunnan,
Southern China, Siam, Malay Peninsula, Sumatra, and Java.

33. ZAMENIS MUCOSUS, L.

Zamenis mucosus, Boul. Cat. Snakes, i. p. 385.

There is a specimen in the British Museum from Singapore from
Dr. Dennys.

Hab. Transcaspia, Afghanistan, India, Ceylon, Burma, Southern
China, Siam, Malay Peninsula, and Java.

34. ZAMENIS FASCIOLATUS, Shaw.

Coluber fasciolatus, Cantor, p. 72.
Zamenis fasciolatus, Boul. Cat. Snakes, i. p. 404.

Cantor obtained this species in Province Wellesley.
Hab. Northern India, Madras, Malay Peninsula.

35. XENELAPHIS HEXAGONOTUS, Cant.

Coluber hexagonotus, Cantor, p. 74.
Ptyas hexagonotus, Stol. J. A. S. B. 1870, p. 186.
Xenelaphis hexagonotus, Boul. Cat. Snakes, ii. p. 8.

I obtained two specimens at Singapore. There are two in the
Raffles Museum labelled Pahang, the largest being 1575 mm. in
length. Cantor obtained one on Penang Hill. Stoliczka found
this species " in a pool of a fresh-water stream on the northern
side of Penang Island," and also in the collection he got from
Penang and Province Wellesley, and there are specimens in the
British Museum from Singapore.

Hab. Burma, Malay Peninsula, Sumatra, Borneo, and Java.

Note.—Peters (Monatsb. Berl. Ac. 1859, p. 269) mentions a
specimen of *Coluber hexagonotus*, Cantor, from Singapore.

36. COLUBER PORPHYRACEUS, Cant.

Psammophis nigrofasciatus, Cantor, P. Z. S. 1839, p. 53.
Coluber porphyraceus, Boul. Cat. Snakes, ii. p. 34.

Cantor obtained a specimen from Singapore.
Hab. Eastern Himalayas, hills of Assam, Burma, Yunnan,
Malay Peninsula, and Sumatra.

37. COLUBER OXYCEPHALUS, Boie.

Herpetodryas oxycephalus, Cantor, p. 80.
Coluber oxycephalus, Boul. Cat. Snakes, ii. p. 56.

Cantor obtained two specimens in the hills of Penang; Stoliczka found it in the collection he got from Penang and Province Wellesley; and there is a specimen in the British Museum from Singapore.

Hab. Eastern Himalayas, Malay Peninsula and Archipelago.

Note.—Peters (Monatsb.. Berl. Ac. 1859, p. 269) mentions a specimen from Malacca.

38. COLUBER MELANURUS, Schl.

Coluber melanurus, Boul. Cat. Snakes, ii. p. 60.

I obtained specimens from Province Wellesley and Singapore; there are several specimens in the Raffles Museum, one being 1830 mm. in length. Stoliczka found it in the collection he got from Penang and Province Wellesley; and there are specimens in the British Museum from Penang and Singapore.

Hab. Southern China, Burma, Malay Peninsula, Sumatra, Borneo, and Java.

39. COLUBER RADIATUS, Schl.

Coluber radiatus, Cantor, p. 73; Boul. Cat. Snakes, ii. p. 61.

Cantor records this species from Penang, Singapore, and the Peninsula; Stoliczka found it in the collection he got from Penang and Province Wellesley; and there is a specimen in the British Museum from Penang from Gen. Hardwicke's collection.

Hab. Southern China, Eastern Himalayas, Bengal, Assam, Burma, Cochinchina, Malay Peninsula, Sumatra, and Java.

40. GONYOPHIS MARGARITATUS, Ptrs.

Gonyophis margaritatus, Boul. Cat. Snakes, ii. p. 71.

Mr. Boulenger mentions a specimen from Singapore, now in the Indian Museum, Calcutta (Ann. Mag. N. H. (6) viii. 1891, p. 290).

Hab. Malay Peninsula, Borneo.

41. DENDROPHIS PICTUS, Boie.

Leptophis pictus, Cantor, p. 82.
Dendrophis pictus, Boul. Cat. Snakes, ii. p. 78.

I obtained this species at Kulim, Kedah, at Taiping, Perak, at Tanglin, Singapore, and from Linga Island. Cantor found it at Penang, and Stoliczka in the collection he got from Penang and Province Wellesley.

Hab. Eastern Himalayas, Bengal, hills of Southern India, Burma, Indo-China, Malay Peninsula and Archipelago.

42. DENDROPHIS FORMOSUS, Boie.

Dendrophis formosus, Boul. Cat. Snakes, ii. p. 84.

I obtained one specimen of this handsome Snake from Crangi, Singapore, and another from Province Wellesley; the latter 1372 mm. in length. The colours of the former when freshly killed were: top of head dark red-brown; upper surface of neck·

red; body and tail bronze-brown, each scale with a distinct black border; a black stripe on each side of the head passing through eye; upper lip, chin, and throat bright citron-yellow; lower parts olive-green, black lines on the lateral keels and each subcaudal distinctly bordered with black. In the latter specimen there were no black lines on the lateral keels, or black borders to subcaudal scales. There is one specimen in the British Museum from Malacca.

Hab. Malay Peninsula, Borneo, and Java.

43. DENDRELAPHIS CAUDOLINEATUS, Gray.

Leptophis caudalineatus, Cantor, p. 85.
Dendrelaphis caudolineatus, Boul. Cat. Snakes, ii. p. 89.

I obtained one specimen at Singapore; there is a specimen in the Raffles Museum labelled Pahang. Cantor mentions it from Penang Hill and Singapore; Stoliczka caught it at Penang, and also found it in the collection he got from Penang and Province Wellesley; and there are two specimens in the British Museum from Singapore from Dr. Dennys, and one from Perak from Mr. Leech.

Hab. Southern India, Mergui, Malay Peninsula and Archipelago.

44. SIMOTES PURPURASCENS, Schl.

Xenodon purpurascens, Cantor, p. 67.
Simotes catenifer, Stol. J. A. S. B. 1873, p. 121, pl. xi. fig. 3.
Simotes dennysi, Blanford, P. Z. S. 1881, p. 218, pl. xxi. fig. 1.
Simotes purpurascens, Boul. Cat. Snakes, ii. p. 218.

Cantor met with this species on Penang Hill; Stoliczka found it in the collection he got from Penang and Province Wellesley, and also records a specimen from Johore; there is a specimen in the British Museum from Singapore; and I obtained one specimen from Province Wellesley.

Hab. South China, Cochinchina, Siam, Malay Peninsula, Sumatra, Borneo, and Java.

45. SIMOTES CYCLURUS, Cant.

Simotes bicatenatus, Stol. J. A. S. B. 1873, p. 114.
Simotes cyclurus, Boul. Cat. Snakes, ii. p. 219.

Stoliczka found this species in the collection he got from Penang and Province Wellesley.

Hab. Bengal, Assam, Burma, Siam, Cochinchina, Southern China, Malay Peninsula, and Sumatra.

46. SIMOTES OCTOLINEATUS, Schn.

Simotes octolineatus, Boul. Cat. Snakes, ii. p. 224.

I obtained a specimen near Taiping, Perak. Ventrals 158; subcaudals 47. The colour was—above very dark brown, beneath white, vertebral line scarlet and three white lines along each side. A specimen caught at Tanglin, Singapore (ventrals 183; subcaudals 61), was coloured yellow, with eight black longitudinal lines, and the space ou centre of back between the two broadest black lines was red ; the belly was also bright red.

A third specimen was caught in the Botanical Gardens, Singapore. Ventrals 168 ; subcaudals 59.   Colours (from spirit) :— Pale brown above, shading to buff underneath, with eight very dark-brown longitudinal lines, those nearest the centre of the back are the broadest, blackest, and most distinctly defined ; the outer lines are narrow, light, and indistinct ; the intermediate rows are transitional in width, colour, and sharpness of outline.

There is a specimen in the British Museum from Singapore from Dr. Dennys.

Hab. Southern India, Malay Peninsula and Archipelago.

47. SIMOTES SIGNATUS, Gthr.

Simotes signatus, Güuth. Rept. Brit. Ind. p. 215, pl. xx. fig. F ; Boul. Cat. Snakes, ii. p. 226.

There are two specimens in the British Museum from Singapore.
Hab. Malay Peninsula, Sumatra, Java.

48. SIMOTES CRUENTATUS, Gthr.

Simotes cruentatus, Stol. J. A. S. B. 1873, p. 121 ; Boul. Cat. Snakes, ii. p. 231, pl. x. fig. 1.

Stoliczka mentions this species as being in the collection he got from Penang and Province Wellesley.
Hab. Burma and Malay Peninsula.

49. ABLABES TRICOLOR, Schl.

Ablabes tricolor, Boul. Cat. Snakes, ii. p. 281.

Mr. Ridley has found this species in the Botanical Gardens at Singapore ; two specimens collected by him there are in the British Museum.
Hab. Malay Peninsula, Sumatra, Borneo, Java.

50. ABLABES BALIODIRUS, Boie.

Coronella baliodeira, Cantor, p. 66.
Ablabes baliodirus, Boul. Cat. Snakes, ii. p. 283.

Cantor obtained two specimens from the hills of Penang ; I obtained one from Province Wellesley.
Hab. Malay Peninsula, Sumatra, Borneo, Java.

51. ABLABES LONGICAUDA, Ptrs.

Ablabes longicauda, Boul. Cat. Snakes, ii. p. 284.

Müller mentions a specimen from Penang in the Bâle Museum (Verh. natur. Ges. Basel, 1882, p. 143).
Hab. Malay Peninsula, Sumatra, Borneo.

52. MACROCALAMUS LATERALIS, Gthr.

Macrocalamus lateralis, Boul. Cat. Snakes, ii. p. 327.

"The only specimen known is from General Hardwicke's East

India collection, and is probably from the Continent" (Günther,
Rept. Brit. Ind. p. 199, pl. xviii. fig. D).
Hab. Malay Peninsula?

53. PSEUDORHABDIUM LONGICEPS, Cant.

Calamaria longiceps, Cantor, p. 63, pl. xl. fig. 1.
Oxycalamus longiceps, Günter, R. B. I. p. 199.
Pseudorhabdium longiceps, Boul. Cat. Snakes, ii. p. 329.

The type specimen was caught on Penang Hill and is preserved
in Dr. Cantor's collection. Stoliczka found one specimen in the
collection he got from Penang and Province Wellesley. There are
specimens in the British Museum from Perak and from Singapore.

I obtained two specimens from Singapore.
Hab. Malay Peninsula and Archipelago.

54. CALAMARIA ALBIVENTER, Gray.

Calamaria linnæi, var., Cantor, p. 62.
Calamaria albiventer, Boul. Cat. Snakes, ii. p. 336.

Cantor records this species from the hills of Penang; there are
also specimens from Penang in Gen. Hardwicke's collection in the
British Museum; and I obtained one specimen from Province
Wellesley.
Hab. Malay Peninsula.

55. CALAMARIA SUMATRANA, Edel.

Calamaria sumatrana, W. L. Sclater, J. A. S. B. lx. 1891,
p. 233; Boul. Cat. Snakes, ii. p. 339.

There is a specimen from Singapore in the India Museum,
Calcutta.
Hab. Malay Peninsula, Sumatra.

56. CALAMARIA LEUCOCEPHALA, D. & B.

Calamaria lumbricoidea, var., Cantor, p. 61.
Calamaria leucocephala, Boul. Cat. Snakes, ii. p. 344.

Cantor records this species from the hills of Penang and
Singapore, and there is a specimen in the British Museum from
Singapore from Dr. Dennys.
Hab. Malay Peninsula, Sumatra, Borneo, Java.

57. CALAMARIA PAVIMENTATA, D. & B.

Calamaria pavimentata, Boul. Cat. Snakes, ii. p. 348.

This Snake appears not to have been recorded before from the
Malay Peninsula. I found one under a stone on Penang Hill,
2100 feet. Ventrals 154; subcaudals 9; length 190 mm.

Colour—upper parts reddish-brown in front, turning to dark
olive-brown on the back, with nine longitudinal black lines; sides of
head yellow, orange collar-mark; bright yellow marks at base and
tip of tail; lower parts greenish-yellow.

I obtained a second specimen from the Province Wellesley; in coloration and marking identical with the Penang specimen.

Hab. Burma, Siam, Cochinchina, Canton, Malay Peninsula, and Java.

Series Opisthoglypha.

Subfamily HOMALOPSINÆ.

58. HYPSIRHINA INDICA, Gray.

Hypsirhina indica, Boul. Cat. Snakes, iii. p. 4, pl. i. fig. 1.

There are two specimens from General Hardwicke's collection in the British Museum, supposed to be from the Malay Peninsula.

Hab. Malay Peninsula?

59. HYPSIRHINA PLUMBEA, Boie.

Homalopsis plumbea, Cantor, p. 101.
Hypsirhina plumbea, Boul. Cat. Snakes, iii. p. 5.

Cantor mentions " two specimens taken in rivulets in the valley of Penang." There is a specimen in the British Museum from Penang from Gen. Hardwicke. Stoliczka found it in the collection he got from Penang and Province Wellesley.

Hab. Burma, Southern China, Indo-China, Malay Peninsula and Archipelago.

60. HYPSIRHINA ENHYDRIS, Schn.

Homalopsis enhydris, Cantor, p. 99.
Hypsirhina enhydris, Boul. Cat. Snakes, iii. p. 7.

There are specimens in the British Museum from Penang from Dr. Cantor, and from Singapore from Mr. Swinhoe. Stoliczka found it in the collection he got from Penang and Province Wellesley.

Hab. India, Ceylon, Burma, Southern China, Cochinchina, Siam, Malay Peninsula and Archipelago.

61. HYPSIRHINA SIEBOLDII, Schl.

Homalopsis sieboldii, Cantor, p. 98.
Hypsirhina sieboldii, Boul. Cat. Snakes, iii. p. 11.

Cantor obtained one specimen from Province Wellesley.

Hab. India, Burma, Malay Peninsula.

62. HOMALOPSIS BUCCATA, L.

Homalopsis buccata, Cantor, p. 96; Boul. Cat. Snakes, iii. p. 14 (skull fig.).

Cantor records this species from Penang and the Peninsula. Stoliczka found it in the collection he got from Penang and Province Wellesley, and there are specimens in the British Museum from Malacca and Singapore.

I obtained two specimens at Singapore; the larger was 1130 mm. in length.

Hab. Burma, Indo-China, Malay Peninsula, Sumatra, Borneo, and Java.

63. CERBERUS RHYNOHOPS, Schn.

Homalopsis rhynchops, Cantor, p. 94.
Cerberus rhynchops, Boul. Cat. Snakes, iii. p. 16.

Cantor mentions this species from the "Malay Peninsula and Islands," and there are specimens in the British Museum from Penang from him, and from Singapore from Dr. Dennys. Stoliczka found it in the collection he got from Penang and Province Wellesley.

This appears to be a common species. I obtained one specimen from Tanglin, Singapore, six from Changi, Singapore (sea-water), and three from Linga Island (sea-water). Seven of these had 23 rows of scales, and three 25 rows; the ventrals varied from 139 to 150 and the subcaudals from 54 to 64; they varied in length from 470 to 670 mm.

Both *Homalopsis* and *Cerberus* seem sluggish on land, and gentle when handled.

Hab. India, Ceylon, Burma, Indo-China, Malay Peninsula and Archipelago, and the Pelew Islands.

64. FORDONIA LEUCOBALIA, Schl.

Homalopsis leucobalia, Cantor, p. 102.
Fordonia leucobalia, Boul. Cat. Snakes, iii. p. 21.

Cantor says this species is found in freshwater, in estuaries, and at sea at Penang and in the Peninsula.

Hab. Rivers and coasts of Bengal, Burma, Cochinchina, Malay Peninsula and Archipelago, New Guinea, and North Australia.

65. CANTORIA VIOLACEA, Gir.

Cantoria elongata, Günther, Rept. Brit. Ind. p. 277.
Cantoria violacea, Boul. Cat. Snakes, vol. iii. p. 23; id. Faun. Brit. Ind., Rept. p. 380 (head figured).

A specimen was procured at Singapore by the U.S. Exploring Expedition, under the command of Capt. Charles Wilkes, U.S.N. (Girard, Proc. Ac. Philadelphia, 1857, p. 182.)

Hab. Burma, Malay Peninsula, Borneo.

66. HIPISTES HYDRINUS, Cant.

Homalopsis hydrina, Cantor, p. 104, pl. xl. fig. 4.
Hipistes hydrinus, Boul. Cat. Snakes, iii. p. 24.

Cantor obtained one specimen from the coast of Penang, and two from the coast of Kedah. There is a specimen in the British Museum from Penang from Mr. Day, and Stoliczka found it in the collection he got from Penang and Province Wellesley. Blanford mentions it from Singapore (P. Z. S. 1881, p. 215).

Hab. Mouths of rivers and coasts of Pegu, Siam, and Malay Peninsula.

Subfamily DIPSADOMORPHINÆ.

67. DIPSADOMORPHUS MULTIMACULATUS, Boie.

Dipsas multimaculata, Cantor, p. 76.
Dipsadomorphus multimaculatus, Boul. Cat. Snakes, iii. p. 63.

Cantor mentions this species from the hills of Penang and the Peninsula.

Hab. Southern China, Indo-China, Burma, Malay Peninsula and Archipelago.

68. DIPSADOMORPHUS GOKOOL, Gray.

Dipsas cynodon, part., Cantor, p. 77.
Dipsadomorphus gokool, Boul. Cat. Snakes, iii. p. 64.

Cantor obtained one specimen on Penang Hill.

Hab. Bengal, Assam, and Malay Peninsula.

69. DIPSADOMORPHUS DENDROPHILUS, Boie.

Dipsas dendrophila, Cantor, p. 76.
Dipsadomorphus dendrophilus, Boul. Cat. Snakes, iii. p. 70.

Cantor records this species from Penang, Singapore, and the Peninsula; Stoliczka found it in the collection he got from Penang and Province Wellesley, and there are specimens in the British Museum from Singapore.

I obtained two large specimens at Kota Star, Kedah; length 1750 and 2310 mm.   The yellow markings were very bright and distinct.

Hab. Malay Peninsula and Archipelago.

70. DIPSADOMORPHUS JASPIDEUS, D. & B.

Dipsadomorphus jaspideus, Boul.Cat. Snakes, iii. p. 73 ; F. Müller, Verh. nat. Ges. Basel, vii. 1882, p. 151.

There is a specimen from Penang in the Bâle Museum.

Hab. Malay Peninsula, Borneo, and Java.

71. DIPSADOMORPHUS DRAPIEZII, Boie.

Dipsadomorphus drapiezii, Boul. Cat. Snakes, iii. p. 74.

There is a specimen in the British Museum from Malacca from Mr. Hervey, also one from Singapore.

I saw one specimen in the jungle on Bukit Tinah, Singapore: ventrals 276, subcaudals 156; length 1524 mm.

Hab. Malay Peninsula and Archipelago.

72. DIPSADOMORPHUS CYNODON, Boie.

Dipsas cynodon, part., Cantor, p. 77.
Dipsadomorphus cynodon, Boul. Cat. Snakes, iii. p. 78.

Cantor obtained this species in Province Wellesley.   Stoliczka mentions *Dipsas cynodon* as being in the collection which he got from Penang and Province Wellesley.   There are specimens in the

British Museum from Malacca from Mr. Hervey, and from
Singapore from Mr. Ridley and Dr. Dennys.
Hab. Assam, Burma, Malay Peninsula and Archipelago.

73. PSAMMODYNASTES PULVERULENTUS, Boie.

Psammodynastes pulverulentus, Boul. Cat. Snakes, iii. p. 172.

Stoliczka mentions this species as occurring in the collection he
got from Penang and Province Wellesley; and there is a specimen
in the British Museum from Kinta, Perak, from Mr. Wray.
Hab. Eastern Himalayas, Khasi and Assam hills, Burma, Indo-
China, Malay Peninsula and Archipelago.

74. DRYOPHIS XANTHOZONA, Boie.

Dryinus prasinus, var. A, Cantor, p. 82.
Dryophis xanthozona, Boul. Cat. Snakes, iii. p. 180.

There is a specimen from Penang from Dr. Cantor in the
British Museum.
Hab. Malay Peninsula and Java.

75. DRYOPHIS PRASINUS, Boie.

Dryinus prasinus, part., Cantor, p. 81.
Dryophis prasinus, Boul. Cat. Snakes, iii. p. 180.

There are specimens in the British Museum from Penang from
Dr. Cantor, and from Singapore from Dr. Dennys. Stoliczka
found it in the collection he got from Penang and Province
Wellesley. I obtained one specimen in Penang; but in Singapore,
about Tanglin, found this Snake in abundance, coloured either
bright green or light brown; judging from the specimens I
observed, the green variety seems to predominate and to grow to
a larger size than the brown. The longest green one I measured
was 1778 mm. in length, but I have seen one about 2000 mm.
These Snakes are very gentle when handled.
Hab. Eastern Himalayas, Assam, Burma, Indo-China, Malay
Peninsula and Archipelago.

76. DRYOPHIOPS RUBESCENS, Gray.

Chrysopelea rubescens, Stoliczka, J. A. S. B. 1870, p. 195.
Dryophiops rubescens, Boul. Cat. Snakes, iii. p. 194.

Stoliczka found one specimen on Penang Hill, and also in the
collection he got from Penang and Province Wellesley.
Hab. Siam, Malay Peninsula and Archipelago.

77. CHRYSOPELEA ORNATA, Shaw.

Leptophis ornatus, part., Cantor, p. 87.
Chrysopelea ornata, Boul. Cat. Snakes, iii. p. 196.

Var. A. There are specimens in the British Museum from
Penang from Dr. Cantor, and from Singapore from Dr. Dennys.
Stoliczka describes this species as common on Penang Hill in 1869,
and found it in the collection he got from Penang and Province

Wellesley. 1 obtained one specimen from Kulim, Kedah, and two from Singapore, the largest being 12:35 mm. in length.

Hab. India, Ceylon, Burma, Southern China, Indo-China, Malay Peninsula and Archipelago.

78. CHRYSOPELEA CHRYSOCHLORA, Reinw.

Leptophis ornatus, part., Cantor, p. 87.
Chrysopelea chrysochlora, Boul. Cat. Snakes, iii. p. 198.

Cantor gives Penang and the Peninsula as the localities from which he obtained this species, and there is a specimen in the British Museum from Singapore from Dr. Dennys.

Hab. Burma, Malay Peninsula and Archipelago.

Series Proteroglypha.

Subfamily HYDROPHIINÆ.

79. HYDRUS PLATURUS, L.

Hydrus bicolor, Cantor, p. 135.
Hydrus platurus, Boul. Cat. Snakes, iii. p. 267.

Cantor obtained a single specimen from the coast of Province Wellesley. Blanford mentions this species (*Pelamis bicolor*) from Singapore, P. Z. S. 1881, p. 215.

Hab. Indian Ocean, Tropical and Subtropical Pacific.

80. HYDROPHIS CÆRULESCENS, Shaw.

Hydrophis cærulescens, Boul. Cat. Snakes, iii. p. 275.

There is a specimen in the British Museum from Penang from Dr. Cantor.

Hab. Bombay coast, Bay of Bengal, and Straits of Malacca.

81. HYDROPHIS NIGROCINCTUS, Daud.

Hydrophis nigrocinctus, Boul. Cat. Snakes, iii. p. 277.

Mr. Boulenger informs me that a specimen in the British Museum from Dr. Blecker is probably from off the coast of Sumatra.

Hab. Bay of Bengal and Straits of Malacca.

82. HYDROPHIS CANTORIS, Gthr.

Hydrus gracilis, part., Cantor, p. 130.
Hydrophis cantoris, Boul. Cat. Snakes, iii. p. 281, pl. xiv.

There is a specimen in the British Museum from Penang from Dr. Cantor.

Hab. Bay of Bengal and Straits of Malacca.

83. HYDROPHIS FASCIATUS, Schn.

Hydrophis fasciatus, Boul. Cat. Snakes, iii. p. 281.

There is a specimen in the British Museum from Penang from Cantor.

Hab. From the coasts of India to China and New Guinea.

84. HYDROPHIS TORQUATUS, Gthr.

Hydrus nigrocinctus, Cantor, p. 128.
Hydrophis torquatus, Boul. Cat. Snakes, iii. p. 283.

Cantor gives as locality of this species "sea of Malayan Peninsula, Penang, and Singapore."
Hab. Bay of Bengal and Straits of Malacca.

85. DISTIRA STOKESII, Gray.

Distira stokesii, Boul. Cat. Snakes, iii. p. 288 (skull fig. p. 286).

There are specimens in the British Museum from Singapore. Blanford mentions two specimens (one 1626 mm. long) from Singapore (P. Z. S. 1881, p. 215).
Hab. Indian Ocean, Straits of Malacca, and north coast of Australia.

86. DISTIRA BRUGMANSII, Boie.

Hydrus striatus, part., Cantor, p. 126.
Distira brugmansii, Boul. Cat. Snakes, iii. p. 292.

There is a specimen in the British Museum from Penang from Dr. Cantor.
Hab. Persian Gulf, coasts of India and Burma, Straits of Malacca, and the Malay Archipelago.

87. DISTIRA CYANOCINCTA, Daud.

Hydrus striatus, part., Cantor, p. 126.
Distira cyanocincta, Boul. Cat. Snakes, iii. p. 294.

There is a specimen in the British Museum from Singapore.
Hab. From the Persian Gulf and the coasts of India to China, Japan, and Papuasia.

88. DISTIRA JERDONII, Gray.

Hydrus nigrocinctus, var., Cantor, p. 129, pl. xl. fig. 8.
Distira jerdonii, Boul. Cat. Snakes, iii. p. 299.

There is a specimen in the British Museum from Penang from Dr. Cantor.
Hab. Bay of Bengal, Straits of Malacca, and Borneo.

89. ENHYDRIS HARDWICKII, Gray.

Hydrus pelamidoides, Cantor, p. 133.
Enhydris hardwickii, Boul. Cat. Snakes, iii. p. 301.

Cantor mentions "sea of Malayan Peninsula and Islands" among the localities of this species. Günther (Rept. Brit. India, p. 380) says of the typical specimen of *Hydrophis hardwickii*, "several circumstances lead me to suppose it was procured at Penang." There are two specimens in the British Museum from Singapore from Mr. Swinhoe.
Hab. From the Bay of Bengal to the Chinese Sea and New Guinea.

90. ENHYDRINA VALAKADIEN, Boie.

Hydrus schistosus, Cantor, p. 132.
Enhydrina valakadien, Boul. Cat. Snakes, iii. p. 302.
There is a specimen in the British Museum from Penang from
Dr. Cantor.
Hab. From the Persian Gulf, along the coasts of India and Burma
to the Malay Archipelago and Papuasia.

91. AIPYSURUS EYDOUXI, Gray.

Aipysurus eydouxii, Boul. Cat. Snakes, iii. p. 304.
Boettger mentions three specimens said to have been caught at
Singapore (Zool. Anz., 1892, p. 420).
I obtained one specimen, a male, from Sourabaya, Java, and
kept it alive in a tin of sea water for about a month, when it died
through an accident. It was gentle when handled, never attempt-
ing to bite. It could move fast, but awkwardly, on dry land, and
sometimes would crawl out of the water of its own accord. The
colours in life are very handsome—above dark olive-brown, with
bright yellow transverse stripes, the stripes and edges of the brown
scales outlined in black; beneath bright yellow. Ventrals 134.
Length 559 mm.
Hab. Seas of Malay Archipelago.

92. PLATURUS COLUBRINUS, Schn.

Laticauda scutata, Cantor, p. 125.
Platurus colubrinus, Boul. Cat. Snakes, iii. p. 308 (skull fig.
p. 307).
There is a specimen in the British Museum from Penang from
Dr. Cantor. Blanford mentions this species (*Platurus scutatus*)
from Singapore (P. Z. S. 1881, p. 215).
Hab. From the Bay of Bengal to the China Sea and the West
South Pacific.

Subfamily ELAPINÆ.

93. BUNGARUS FASCIATUS, Schn.

Bungarus fasciatus, Cantor, p. 113; Boul. Cat. Snakes, iii.
p. 366.
Cantor mentions this species from Penang and Prov. Wellesley,
and Stoliczka found it in the collection he got from Penang and
Province Wellesley. Blanford mentions it from Singapore.
Hab. India, Burma, Southern China, Indo-China, Malay Peninsula, Sumatra, and Java.

94. BUNGARUS CANDIDUS, L.

Bungarus candidus, Cantor, p. 113; Boul. Cat. Snakes, iii.
p. 368 (skull fig. p. 365).
Cantor mentions this species from Kedah, and there are five

specimens from him in the British Museum from Penang and the Peninsula.

Hab. India, Burma, Southern China, Indo-China, Malay Peninsula, Java, and Celebes.

95. BUNGARUS FLAVICEPS, Reinh.

Bungarus flaviceps, Cantor, p. 112; Boul. Cat. Snakes, iii. p. 371.

I obtained one specimen from Province Wellesley. Ventrals 237; subcaudals 53, of which the first 16 were single and the remainder double, except the 19th, 29th, 30th, 31st, and 32nd. There were three postoculars on the right side. It was 1473 mm. in length.

Cantor mentions obtaining one specimen on Penang Hill.

Hab. Tenasserim, Cochinchina, Malay Peninsula, Sumatra, Borneo, and Java.

96. NAIA TRIPUDIANS, Merr.

Naja lutescens, Cantor, p. 117.
Naia tripudians, Boul. Cat. Snakes, iii. p. 380.

Cantor says this species is found in Penang, Singapore, and the Peninsula, and that the brown variety prevails at Penang and the black at Singapore. Several residents in the Settlements have told me the same thing. The largest Cobra I met with was a black one in Singapore, 1372 mm. long. Mr. Ridley caught in the Botanical Gardens, Singapore, a Cobra in the act of swallowing a *Macropisthodon rhodomelas.* A Cobra that I obtained from Kulim, Kedah, belonged to a third colour variety, C. *b.* in Boulenger's 'Catalogue of Snakes.'

Hab. Southern Asia, from Transcaspia to China and the Malay Archipelago.

97. NAIA BUNGARUS, Schl.

Hamadryas ophiophagus, Cantor, p. 116.
Naia bungarus, Boul. Cat. Snakes, iii. p. 386.

Cantor records this species from Penang Hill and Province Wellesley, and there is a specimen in the British Museum from Singapore from Dr. Dennys. From all accounts the Hamadryad is still common in the hills of Penang, and I have seen several skins of large individuals killed near Taiping, Perak.

Hab. India, Burma, Indo-China, Southern China, Malay Peninsula and Archipelago.

98. CALLOPHIS GRACILIS, Gray.

Elaps nigromaculatus, Cantor, p. 108, pl. xl. fig. 7.
Callophis gracilis, Boul. Cat. Snakes, iii. p. 396.

Cantor records this species from the hills of Penang and from Singapore.

Hab. Malay Peninsula and Sumatra.

99. CALLOPHIS MACULICEPS, Gthr.

Elaps melanurus, Cantor, p. 106, pl. xl. fig. 6.
Callophis maculiceps, Boul. Cat. Snakes, iii. p. 397.
Cantor obtained one specimen from Province Wellesley.
Hab. Cochinchina and Malay Peninsula.

100. DOLIOPHIS BIVIRGATUS, Boie.

Elaps bivirgatus, Cantor, p. 109.
Elaps flaviceps, Cantor, P. Z. S. 1830, p. 33.
Doliophis bivirgatus, Boul. Cat. Snakes, iii. p. 400.

Cantor obtained this species from the hills of Penang and from Malacca; Stoliczka found it in the collection he got from Penang and Province Wellesley, and there are specimens in the British Museum from Penang and Singapore.

I obtained specimens from Kulim, Kedah, from Singapore, and from Province Wellesley, the latter 1372 mm. in length.

Hab. Burma, Cochinchina, Malay Peninsula, Sumatra, Borneo, and Java.

Note.—Girard, Proc. Ac. Philad. 1857, p. 182, records a specimen of *Doliophis flaviceps* from Singapore.

101. DOLIOPHIS INTESTINALIS, Laur.

Elaps intestinalis, Cantor, p. 107.
Elaps furcatus, Cantor, P. Z. S. 1839, p. 34.
Doliophis intestinalis, Boul. Cat. Snakes, iii. p. 401.

Of this Snake Cantor says "it is of no uncommon occurrence in the hills of Penang, at Malacca, and at Singapore." Stoliczka found it in the collection he got from Penang and Province Wellesley. I obtained two specimens from Tanglin, Singapore, and two from Province Wellesley, one of the latter belonging to the variety *trilineatus*. There are in the British Museum specimens of the variety *lineata* from Penang and Singapore, and of *annectens* from Singapore.

Hab. Burma, Malay Peninsula and Archipelago.

Family AMBLYCEPHALIDÆ.

102. HAPLOPELTURA BOA, Boie.

Dipsas boa, Cantor, p. 78, pl. xl. fig. 3.
Haplopeltura boa, Boul. Cat. Snakes, iii. p. 439.
Cantor obtained this species from the hills of Penang.
Hab. Malay Peninsula and Archipelago.

103. AMBLYCEPHALUS LÆVIS, Boie.

Amblycephalus lævis, Boul. Cat. Snakes, iii. p. 441.
This Snake is said to have been found at Malacca, but I have not been able to discover where it is recorded.
Hab. Malay Peninsula, Natuna Islands, Borneo, and Java.

104. AMBLYCEPHALUS MALACCANUS, Ptrs.

Asthenodipsas malaccana, Peters, Mon. Berl. Ac. 1864, p. 273, pl. —. fig. 3.

Amblycephalus malaccanus, Boul. Cat. Snakes, iii. p. 442.

One specimen was obtained in the neighbourhood of Malacca (*see* Peters, *l. s. c.*).

Hab. Malay Peninsula, Sumatra, Borneo.

Family VIPERIDÆ.

Subfamily CROTALINÆ.

105. LACHESIS MONTICOLA, Gthr.

Trimeresurus convictus, Stoliczka, J. A. S. B. 1870, p. 224, pl. xii. fig. 1.

Lachesis monticola, Boul. Cat. Snakes, iii. p. 548.

Stoliczka caught one specimen on Western Hill, Penang, and there is a specimen in the British Museum from Singapore.

Hab. Tibet, Himalayas, Assam, Burma, Malay Peninsula, and Sumatra.

106. LACHESIS PURPUREOMACULATUS, Gray.

Trigonocephalus puniceus, Cantor, p. 122.

Lachesis purpureomaculatus, Boul. Cat. Snakes, iii. p. 553.

Cantor records this species from Penang and the Peninsula.

Var. A.—There are specimens in the British Museum from Penang and Singapore. I obtained one from Tanjong Katong, Singapore. Ventrals 172; subcaudals 57 (double, except the 2nd and 3rd, which were single). Scales in 25 rows.

Var. B.—There is one specimen in the British Museum from Penang from Dr. Cantor.

Hab. Himalayas, Bengal, Assam, Burma, Andamans, Nicobars, Malay Peninsula, and Sumatra.

107. LACHESIS GRAMINEUS, Shaw.

Trigonocephalus gramineus, part., Cantor, p. 119.

Trimeresurus erythrurus, Stoliczka, J. A. S. B. 1870, p. 217.

Lachesis gramineus, Boul. Cat. Snakes, iii. p. 554.

Cantor gives Penang, Singapore, and the Peninsula as localities of this species; Stoliczka obtained specimens from Penang and Province Wellesley; and I obtained four specimens from Province Wellesley. Blanford mentions this species from Singapore.

Hab. South-eastern Asia.

108. LACHESIS SUMATRANUS, Raffles.

Lachesis sumatranus, Boul. Cat. Snakes, iii. p. 557.

There is a specimen in the British Museum from Singapore from Dr. Dennys.

Hab. Malay Peninsula and Archipelago.

109. LACHESIS WAGLERI, Boie.

Trigonocephalus sumatranus, Cantor, p. 121, pl. xl. fig. 9.
Lachesis wagleri, Boul. Cat. Snakes, iii. p. 562.

Cantor gives Penang, Singapore, and the Peninsula as localities of this species; Stoliczka found it in the collection he got from Penang and Province Wellesley. There are specimens in the British Museum from Penang, Taiping (Perak), Malacca, and Singapore. I obtained a specimen on Bukit Timah, Singapore. Blanford mentions this species from Singapore and Selangor.

Hab. Malay Peninsula and Archipelago.

Class BATRACHIA.

Order ECAUDATA.

Suborder PHANEROGLOSSA.

Series FIRMISTERNIA.

Family RANIDÆ.

1. OXYGLOSSUS LIMA, Tschudi.

Oxyglossus lima, Boul. Cat. Batr. Sal. p. 5.

This species is said to occur in the Malay Peninsula, but I have not been able to find it recorded south of Tenasserim, though it occurs again in Java.

Hab. Lower Bengal, Burma, Southern China, Camboja, Siam, Malay Peninsula, Java.

2. OXYGLOSSUS LÆVIS, Gthr.

Oxyglossus lævis, Boul. Cat. Batr. Sal. p. 6.

There are specimens in the British Museum from Perak from Mr. Wray, one from Larut, the other from Changkatjerin. I found two specimens in the Raffles Museum, unlabelled.

Hab. Burma, Malay Peninsula and Archipelago, Philippine Islands.

3. RANA CYANOPHLYCTIS, Schn.

Rana leschenaultii, Cantor, p. 138.
Rana cyanophlyctis, Boul. Cat. Batr. Sal. p. 17.

Cantor mentions two specimens from the Malay Peninsula, and says " the species is apparently not numerous."

Hab. South Arabia, Baluchistan, Cashmere, Himalayas (up to 6000 ft.), India, Ceylon, Malay Peninsula.

4. RANA LATICEPS, Blgr.

Rana laticeps, Boul. Cat. Batr. Sal. p. 20, pl. i. fig. 1.

There is a specimen in the British Museum from Malacca from

58*

Mr. Hervey. A Frog in the Raffles Museum, Singapore, labelled
" Malacca," is apparently of this species, but it is in a bad state of
preservation, and the back is quite smooth, without the tubercles
which are present in Mr. Hervey's specimen ; both are females.

Hab. India, Malay Peninsula.

5. RANA MACRODON, Kuhl.   (Plate XLV. fig. 1.)

Rana fusca, Stol. J. A. S. B. 1873, p. 115.

Rana macrodon, Blanford, P. Z. S. 1881, p. 225, pl. xxi. fig. 4
(upper view of head); Boul. Cat. Batr. Sal. p. 24, pl. i. fig. 4
(inside of mouth).

As first pointed out by Mr. Blanford, there seem to be two varieties of this species, very different in appearance.   The specimens I
collected at Penang are so different from those I got at Singapore,
as to appear to be of distinct species ; but on comparing them
with the large series in the British Museum from many different
localities in the East Indies, I cannot find any constant characters
by which to separate the two varieties.

Stoliczka found this species in the collection he got from Penang
and Province Wellesley, but from his description one cannot tell
to which variety his specimens belonged.   F. Müller mentions a
specimen of *Rana macrodon* (Günth. Cat. Batr. p. 8) from Malacca
in the Bâle Museum (Verh. naturforsch. Ges. Basel, vii. 1882-85).

The following description will, I trust, be of use in identifying
this Frog: provisionally I have called the broad-headed form the
Singapore variety, and the narrower-headed the Penang variety.

Vomerine teeth on two straight ridges running obliquely back
from the anterior angle of the choanæ, and converging behind so as
to meet, if prolonged, nearly in a right angle, but rather widely
separated ; a strong osseous transverse ridge behind the choanæ;
lower jaw with two fang-like bony prominences in front, fitting
into hollows inside the upper jaw ; when the mouth is closed, the
size to which these prominences are developed is variable.

Head large, this is especially so in the adults of the Singapore
variety.   In the typical Penang variety the snout is usually
pointed, but very variable in shape ; in the Singapore variety it is
broad and rounded at the end.   Blanford says of the snout of the
Singapore variety, " no trace of *canthus rostralis,*" but in my specimens, though but slightly developed, it is at once apparent ; no
constant distinction can be made between the two varieties in
regard to the amount of depression of the snout.   Occiput more
or less swollen at the sides.   The nostrils are nearer the end of
snout than the eye ; their distance apart in the Penang variety is
equal to or greater than the interorbital space, while in the Singapore variety it is *considerably* less: this character will be found
useful in distinguishing between the two varieties, but it does not
hold good for young specimens.   In all seven Penang specimens
the breadth across the gape is about equal to the distance from
angle of mouth to end of snout, and considerably less than the

length of the hind foot; while in all the adult Singapore specimens examined the breadth across the gape is greater than the distance from angle of mouth to end of snout, and equal to or greater than the length of the hind foot, but in the young of the Singapore variety the gape is less than the hind foot.   The interorbital space in the largest Penang specimen is equal to, in the six others less than, the upper eyelid, in some considerably less; in the Singapore variety, in young specimens the interorbital space is slightly less than the upper eyelid, in fair-sized specimens equal to it, and in large specimens one half broader than the upper eyelid.   Blanford mentions the Singapore frog as having a smaller eye; but if specimens of similar size of the two varieties are compared, it will be seen not to be noticeable.   Tympanum distinct, slightly larger in the Singapore variety, but variable in size; it is also variable in shape, when not circular, in the Singapore variety it has its greater diameter in a vertical position, in the Penang variety in a horizontal direction.   In the Singapore variety a strong, prominent fold (well-developed in even quite small specimens) runs from behind the eye horizontally to over the tympanum, and then turns down at an obtuse angle and runs straight to the shoulder; in the Penang variety this fold is much less prominent, and instead of forming an obtuse angle forms a curve above the tympanum; however, this character cannot divide the two forms, as in the British Museum specimens will be found with every gradation from the angular to the curved fold.

Fingers moderate, first much longer than the second; toes broadly webbed, in the Singapore variety the web is more deeply emarginate than in the Penang variety, the terminal two phalanges of the fourth toe have only a narrow fringe of web along their sides.   The tarsal fold is very variable in size, and often wanting. The fingers and toes have slightly though distinctly swollen tips, and the subarticular tubercles of fingers and toes are well developed; the inner metatarsal tubercle is elongate and blunt, there is no outer tubercle: in these characters there is no difference between the two varieties.   The hind limb being carried forward along the body the tibio-tarsal articulation reaches beyond the eye, usually to the end of the snout: the Penang specimens have on the whole longer hind legs than those from Singapore when measured in this way.

Skin smooth above.   Hinder portion of upper eyelid tubercular. In young specimens there is a narrow glandular fold on each side of the back, and other, both round and longitudinal, glands scattered over the skin of the upper surfaces; these glands gradually disappear with age, but seem more persistent in the Penang variety.   Male without vocal sacs.

Blanford distinguished the Singapore variety from *Rana fusca* (Blyth) by, 1st, a much broader head; 2nd, a smaller eye; 3rd, a larger tympanum; 4th, flatter muzzle; 5th, nostrils nearer together; 6th, web of the hind toes less developed.   Although, as mentioned before, I can find no constant characters to separate the two varieties, the following points should be noticed :—

(i.) Breadth of interorbital space compared to the distance between the nostrils.

(ii.) General form of the snout.

(iii.) Shape and prominence of the tympanic fold.

(iv.) Shape of the tympanum.

(v.) Amount of emargination in the webbing of the hind toes.

Localities. Of the Penang variety I collected seven specimens in small ponds on Penang Hill, at elevations of from 2000' to 2200', in March 1896. They are active frogs and good swimmers, and locally called "Koldok-ayer" (Malay). There is in the British Museum a specimen from Mr. Wray, from the hills of Larut, Perak, at an elevation of between 3000' and 4000', which agrees with the Penang variety in the more pointed snout, in the distance between the nostrils being greater than the interorbital space, in the skin having longitudinal glandular folds, and in the webbing of the hind feet, but the tympanic fold is angular.

Of the Singapore variety I collected eight specimens from the following places in the island—Passir Panjang, Botanical Gardens, and Bukit Timah, at elevations of less than 400 ft., in January and April 1896. Four large specimens from Dr. Dennys, one from the Raffles Museum, and three young specimens from Mr. Ridley, all from Singapore, and now in the British Museum, agree with my specimens of corresponding sizes, and are distinctly of this variety, the full-grown ones showing well the characteristic broad head and angular prominent tympanic fold. There are several large specimens of this variety in the Raffles Museum, Singapore; one (in spirit) has a *Dryophis prasinus* in its mouth. This Frog seems common but local in Singapore island, and is known as the "red frog" or "Koldok-merah" (Malay): it is a very handsome animal from its athletic build, bright eye, and brilliant colour, which last, however, helps to conceal the frog when (as I have more than once found it) among large fallen leaves of the same bright red as itself. When frightened, both the Penang and Singapore varieties take to the water, diving straight in and seeking concealment immediately at the bottom.

Although, as far as we know, only the one variety inhabits Penang and the other Singapore, there are specimens of both in the British Museum from Java, and also from Borneo, where are also intermediate forms with the angular tympanic fold, but the distance between the nostrils greater than the interorbital space, and with fully-webbed hind feet. There is a specimen in the British Museum from Great Natuna Island, from Mr. Hose, which seems identical with the Singapore variety.

Colour, from life. Specimens from Penang Hill.—Upper parts rich dark olive-brown or green, with or without a broad orange vertebral line. Chin white. Belly and lower side of limbs pale orange.

Specimens from Singapore.—Upper parts bright bronze or chocolate-red, varies very much in intensity, in captivity becomes a pale yellowish- or brownish-red. In one half-grown specimen the upper parts were a dark olive-brown. A very narrow pale yellow vertebral stripe seems usually present, but often very irregular, not following the centre of the back. Lower surfaces yellow, paler or bluish-white on the throat, more or less spotted or mottled with black. Lips very pale yellow, extensively marked with black. A black line under the fold from eye to tympanum, continued but narrower to angle of mouth. Iris golden. Limbs indistinctly barred with dark brown; a narrow, pale yellow, distinct stripe down the hind leg, the skin behind this is white or yellow, marbled with black; the web between the toes is dark brown. Quite small specimens, of about 40 mm. in length, are very differently coloured from the adults, and somewhat resemble *Rana limnocharis.*

Size. The largest Penang specimen is 92 mm. from snout to vent. The largest Singapore specimen I have measured is 105 mm. from snout to vent, and the width of the head at the angle of the mouth is 76 mm. This species seems to attain a larger size in Singapore than in any other locality.

Hab. Upper Burma, Tenasserim, Malay Peninsula and Archipelago.

6. RANA PLICATELLA, Stol.

Rana plicatella, Stol. J. A. S. B. 1873, p. 116, pl. xi. fig. 1; Boul. Cat. Batr. Sal. p. 26.

This Frog was discovered by Stoliczka in the collection he got from Penang and Province Wellesley.

Hab. Malay Peninsula.

7. RANA TIGRINA, Daud.

Rana tigrina, Cantor, p. 139; Boul. Cat. Batr. Sal. p. 26; id. Fauna Brit. Ind., Rept. p. 449 (figured).

Cantor says this species "is excessively numerous in valleys and hills, after heavy falls of rain, Malayan Peninsula and Islands."

Stoliczka (J. A. S. B. 1873, p. 112) mentions *Rana tigrina,* var. *pantherina,* in the collection he got from Penang and Province Wellesley. There are in the British Museum specimens from Penang, from Dr. Cantor, Major Sykes, and Sir A. Smith.

In April 1895 I found this Frog common in the evening about Kota Star, Kedah.

Hab. Nepal, Sikhim, India, Ceylon, Burma, China, Formosa, Siam, Malay Peninsula and Archipelago.

8. RANA LIMNOCHARIS, Boie.

Rana gracilis, Boul. Cat. Batr. Sal. p. 28; Stol. J. A. S. B. 1870, p. 142.

Rana lymnocharis, Stol. J. A. S. B. 1873, p. 116.

Rana limnocharis, Boul. Fauna Brit. Ind., Rept. p. 450.

Stoliczka says this species is very common in Penang and Province Wellesley, and from Penang Hill (2000 ft.) he obtained a variety which he called *pulla*. There are specimens in the British Museum from Perak from Mr. Wray, from the Dindings from Mr. Ridley, and from Malacca from Mr. Hervey. This Frog was common about Tanglin, Singapore; usually, in April, found sitting on the banks of ponds in the evening; it does not attempt to escape by jumping into the water like *Rana tigrina, R. macrodon,* and *R. flammea,* but even if touched squats down close on the clay, which its colour does not resemble, so is easily caught. Stoliczka (J. A. S. B. 1870, p. 153) mentions *Bufo penangensis* as having a similar habit. The largest Tanglin specimen was 62 mm. from snout to vent. Their usual coloration seems, pale olive-green above, with dark green blotches and a distinct, narrow, pale yellow dorsal stripe; the underneath being immaculate buff, except the lips which have distinct black spots, and the throat (male) has two large black blotches.

Hab. Sikhim, India, Ceylon, Burma, China, Formosa, Japan, Siam, Malay Peninsula and Archipelago.

9. RANA HASCHEANA, Stol.

Polypedates hascheanus, Stol. J. A. S. B. 1870, p. 147, pl. ix. fig. 3.

Rana hascheana, Sclater f., P. Z. S. 1892, p. 344.

Stoliczka says: " I found this species tolerably common all through the higher forests (about 1000 feet above sea-level) in the island of Penang ; . . . . . . I have seen hundreds of specimens in different places of the island, ... It is generally seen on the leaves of small bushes or on the ground between old leaves."

Hab. Malay Peninsula and Natuna Islands.

10. RANA ERYTHRÆA, Schl. (Plate XLV. fig. 2.)

Limnodytes erythræus, Cantor, p. 141.
Hylarana erythræa, Günther, Rept. Brit. Ind. p. 425.
Rana erythræa, Boul. Cat. Batr. Sal. p. 65.

Cantor mentions having observed three individuals from the Malay Peninsula. Stoliczka (J. A. S. B. 1873, p. 112) found it in the collection he got from Penang and Province Wellesley. There is a specimen in the British Museum from Perak from Mr. Wray. I found one individual in the Lines, Penang, in May 1895, but in Singapore in April 1896. I found it excessively numerous about the ponds at Tanglin and in the Botanical Gardens, in ditches near Thompson Road and in the low-lying fields up the Singapore river, where it may be heard croaking at night. This is a most active, agile Frog, both on land and in the water; it can hop over the surface of a pond, much as *Rana cyanophlyctis* does in India, and also jump right out of the water. Owing to the vivid green colour of its back exactly matching the colour of the weeds in a pond, it is often difficult to see but for its

bright golden eyes. The largest specimens were 72 mm. in length
from snout to vent.

Colour (from life).—Above the most vivid green, exactly
matching some of the water-weeds in ponds, but in other surroundings the back may change to a dull green or a yellowish
brown : no specimens that I met with had "back and sides brown
or reddish olive " as described by Cantor, from life? A very dark
brown stripe (generally darker at the edges) runs along each side
of the head and body from the nose to the inset of the hind leg
(in one specimen these side stripes were bright green, like the
back, with black edges); this broad dark stripe is separated from
the green back by a distinct white or yellowish-white stripe. The
upper lip is yellow. The limbs are reddish-buff or yellowish-
brown, paler beneath. The underneath of the head and body is
immaculate, pure white. The iris is golden or golden-orange.

Hab. Burma, Siam, Malay Peninsula and Archipelago.

11. RANA LABIALIS, Blgr. (Plate XLV. fig. 3.)

Rana labialis, Boul. Ann. & Mag. N. H. 1887, (5) xix. p. 345,
pl. x. fig. 1.

This Frog was described from several specimens from Malacca
given to the British Museum by Mr. Hervey ; specimens have
since been received there from Singapore from Mr. Ridley. I
caught two specimens at Tanglin, Singapore, in a small pond on
the 2nd of April, 1896; it appeared fairly numerous, and was
associated with *Rana erythræa*, which it resembles in colour,
having the upper parts bright green and the lower immaculate
white : this bright green in spirit becomes dull and dark.

Hab. Malay Peninsula and Mentavi Islands.

Tadpole.—I found tadpoles of this species in a small pond in
the Botanical Gardens, Singapore, in the middle of April 1896.
Length of body about once and a half its width, about two-thirds
the length of the tail. Nostrils, as seen from above, nearer the
end of the snout than the eyes. Eyes on the upper surface of the
body, rather nearer the end of the snout than the spiraculum ; the
distance between the eyes twice as great as that between the
nostrils, and greater than the width of the mouth. Spiraculum
on the left side, directed upwards and backwards, situated nearer the
anus than the end of the snout, visible from above and from below.
Anus opening on the right side, close to the lower edge of the
subcaudal crest. Tail three to four times as long as deep, ends in
a rounded point, intermediate in shape between those of *Rana
esculenta* and *Rana temporaria* (Boul. P. Z. S. 1891, pl. xlv. figs. 1,
3); upper crest convex, slightly deeper than the lower, not
extending on to the back ; the depth of the muscular portion, at
its base, about half or rather more of its greatest total depth.

Beak edged with black. Sides and lower edge of the lip fringed
with papillæ, those on the lower edge being long and prominent ;
upper lip with four series of fine teeth, the outermost is uninter-

rupted, the three inner broadly interrupted and very short, decreasing in size towards the mouth, the innermost row is sometimes very small and inconspicuous; lower lip with three series of teeth, the two outer uninterrupted, the third narrowly interrupted, the three rows are about the same length, but the median is the longest and the outermost the shortest.

The colour of these tadpoles in life is brick-red above, and pale yellow beneath, but the whole skin is very transparent, the eyes and the intestines being clearly seen. On the back on each side behind the eyes is a patch of granulated skin; in some specimens there is a similar strip on the hinder part of the back on each side parallel with the tail, and a large patch on each side of the belly, oblong in shape, and each converging together towards the tail.

A good specimen measures 37 mm. in total length; body 15; width of body 10; tail 22; depth of tail 6.

Depth of muscular portion of tail at its base between 3 and 4 mm.

The above measurements are taken from a spirit-specimen.

12. RANA LUCTUOSA, Ptrs. (Plate XLVI.)

Rana luctuosa, Boul. Cat. Batr. Sal. p. 68.

This handsome little Frog appears to have hitherto only been recorded from Borneo; in March 1896 I found it common about certain small ponds on Penang Hill at an elevation of 2000 feet. They were generally in long grass near the water's edge; when alarmed they would jump into the water, but before long crawl out again.

Colour (from life).—Top of head and back rich dark chocolate-brown (in very small frogs of this species the back is a very bright red, more vermilion than chocolate), bordered on each side from the nose to the insertion of the hind leg by a very distinct white line. Sides of head, neck, and body are very dark brown or black. The tympanum is dark reddish-brown. Along the lower part of the sides of the body are a few white spots in an irregular line from angle of mouth to thigh. Lower surfaces—chin and throat dark brown, remainder dirty buff, darker on limbs. Limbs very dark brown or bluish-black, with bluish-white or very pale grey marblings; the black turns to brown on the toes, and the marbling is also less conspicuous on the feet.

These colours seem permanent and not variable according to surroundings, as is the case with many batrachians.

Adult specimens are 45 to 50 mm. from snout to vent.

Hab. Malay Peninsula; Borneo.

Tadpole, "Koldok-ikan" (Malay). In March 1896, in a pond of clear water (2200 ft. elevation) in the jungle on Penang Hill, there were a large number of tadpoles of this species and little frogs just leaving the water.

Description of the Tadpole.

Length of body once and a half its width, considerably more than half the length of the tail. Nostrils nearer the end of the

snout than the eye. Eyes on the upper surface of the body,
nearer the end of the snout than the spiraculum; a well-marked
lachrymal gland from the eye to the nostril; the distance between
the eyes twice as great, or rather more, than the distance between
the nostrils and much greater than the width of the mouth.
Spiraculum on the left side, directed backwards and upwards,
nearer the anus than the end of the snout, visible from above and
from below. Anus opening on the right side, close to the lower
edge of the subcaudal crest. Tail between three and four times as
long as deep, acutely pointed, the tip being inclined upwards in
life; upper crest convex, about equal in depth to the lower; the
upper crest does not extend on to the back. Depth of the
muscular portion of the tail at its base rather more than half
greatest total depth of tail.

Beak broadly edged with black. Sides and lower edge of the lip
bordered with papillæ. Upper lip with six series of fine teeth,
the upper uninterrupted, the remainder broadly interrupted,
decreasing in length towards the beak, the sixth or inner series in
some specimens being very small or absent. Lower lip with four
long series of teeth, the inner very narrowly interrupted, the
remainder uninterrupted.

Colour. The larva till it reaches a total length of from 35 to
40 mm. is blackish-brown above, white beneath, with a grey
mottled tail; after this the upper parts are a warm-brown,
mottled with darker brown, and the sides and lower parts yellow
also mottled with brown, but the skin of the underneath of the
abdomen is transparent and of a purple colour; the tail is mottled
brown and yellow; the hind legs of the tadpole are grey marbled
with black, when the fore legs appear the back assumes the bright
chocolate colour of the adult frog. The iris is golden, and the eye
bright and noticeable.

Size. Length of body 24 mm. Length of tail 44 mm. Depth
of tail 12 mm. The largest tadpole, without hind legs, I observed
was 70 mm. in total length; the largest, with hind legs but
without fore legs, was 77 mm. The recently transformed young
measure about 25 mm. from snout to vent.

13. RANA GLANDULOSA, Blgr.

Rana glandulosa, Boul. Cat. Batr. Sal. p. 73, pl. vii.

There are specimens in the British Museum from Malacca from
Mr. Hervey and from Singapore from Mr. Ridley.

Hab. Malay Peninsula, Borneo, and Palawan.

14. RHACOPHORUS LEUCOMYSTAX, Gravh.   (Plate XLIV. fig. 2.)

Polypedates leucomystax, Cantor, p. 142.

Polypedates maculatus, Stol. J. A. S. B. 1870, p. 148, & 1873,
p. 112.

Rhacophorus maculatus, part., Boul. Cat. Batr. Sal. p. 83.

Rhacophorus leucomystax, Boul. Faun. Brit. Ind., Rept. & Batr.
p. 474.

Cantor gives Penang, Singapore, and the Malay Peninsula as localities, and says "although it inhabits Singapore . . . . . it appears not to occur in the valleys of Penang, but to affect the hills, at an elevation of more than 2000 ft." Stoliczka mentions it as being " not uncommon in Penang." I found this species very common both in Penang and Singapore, but, contrary to Cantor's experience, I found it at Penang at almost the sea-level (20 ft.), though it was certainly more numerous on the hills. It is a cheerful little frog of most graceful build. It comes out from its hiding-places shortly before sunset, and remains abroad all night; the males are easily found as they sit on shrubs or trees or on the edges of the rainwater-butts under the verandahs of the houses, and from time to time utter a single rather musical short croak. In March and April they can be found both by day and night *in copulâ* in ponds. Cantor mentions the power of changing its colours this species possesses. It changes both its colour and markings very rapidly and frequently, but dark bands across the legs can always be more or less distinguished; the lower parts are some shade or other of buff, but the principal variations of the upper parts are as follows :—

      (i.) pale bronze, uniform ;
      (ii.) pale bronze, with four longitudinal dark brown or black lines ;
      (iii.) a bright yellowish-bronze, almost orange, uniform ;
      (iv.) reddish-brown, almost chocolate, mottled with darker ;
      (v.) pale brownish-green or olive, with irregular dark spots ;
      (vi.) yellowish-green, mottled with darker or brown.

The *Rhacophorus* mentioned by Stoliczka (J. A. S. B. 1873, p. 112) as a separate species, *Polypedates quadrilineatus*, from Penang, is not even a true variety, as the dark lines appear conspicuously and disappear entirely in the same individual. If killed with or without the lines visible they remain so in spirit. In Singapore at different times I noticed many young frogs which had just left the water all of which had the dark lines visible; these disappear as the animal grows, only to reappear temporarily in the adult.

The females are considerably larger than the males ; the largest male I caught was 48 mm. from snout to vent, and the largest female 68 mm. from snout to vent.

Hab. Sikhim, Assam, Burma, Southern China, Malay Peninsula and Archipelago, Philippines.

Tadpole. In January, February, March, and April, 1896, I found the tadpoles of this species in several small ponds and in rainwater-butts about Singapore; and was able to collect a large series for the British Museum.

Description of the Tadpole.

Length of body once and a half its width, half the length of the

tail or rather less. Nostrils nearer the end of the snout than the
eyes. A strongly-marked lachrymal gland from eye to nostril.
Eyes on the side of the head, nearer the spiraculum than the end
of the snout; the distance between the eyes more than twice as
great as that between the nostrils, and much greater than the
width of the mouth.

Spiraculum on the left side, directed backwards and upwards,
nearer the anus than the end of the snout, visible from above and
from below. Anus opening on the right side, halfway between
the lower edge of the subcaudal crest and the muscular portion of
the tail. Tail rather more than three times as long as deep, very
acutely pointed, upper crest convex, about the same depth as the
lower, or in some specimens very markedly shallower; the upper
crest does not extend on to the back; the depth of the muscular
portion at its base rather more than half the greatest total length.

Beak black. Sides and lower edge of the lip bordered with
papillæ, except in the centre of the lower lip, where there is a
small semicircular notch, devoid of papillæ. Upper lip with four
series of fine teeth, the uppermost uninterrupted, the second
narrowly interrupted, and the third and fourth broadly so; lower
lip with three long uninterrupted series of teeth.

Colour. Above dark brown, irregularly mottled with darker;
beneath buff; the sides and tail buff, mottled with brown. These
tadpoles, from different localities, vary a good deal in colour, some
being dark brown above, others a light dirty buff colour.

Size. These tadpoles vary even more in size than in colour;
some exceptionally fine ones were 46 mm. in total length. Length
of body 15·5 mm., length of tail 31, depth of tail 10.

The recently transformed young measure from 14 to 18 mm.
from snout to vent.

15. RHACOPHORUS LEPROSUS, Schl.

Polypedates leprosus, Günther, Ann. & Mag. N. H. (5) xx. 1887,
p. 315, pl. xvi. figs. A, *a*, *a'*.
Rhacophorus leprosus, Boul. P. Z. S. 1890, p. 284.

Mr. Wray obtained this species at an elevation of 4000 ft. on
the hills of Larut, Perak. He says of it:—"This species . . . . . .
lives in holes in trees, and the note produced by it is not so loud
as that of *Phrynella*, and has a more metallic ring in it."

Hab. Malay Peninsula; Sumatra.

NOTE.—*Rhacophorus dennysi* was described by Mr. Blanford
(P. Z. S. 1881, p. 224, pl. xxi. fig. 3); the specimen was in a
collection sent from Singapore by Dr. Dennys, and was said to
have come from China. Since then another specimen of this species
has been received at the British Museum from Foochow; so that
there can be little doubt that the type specimen was really from
China, and that this species should not be included in the fauna
of Malaya.

16. IXALUS PICTUS, Ptrs.

Ixalus pictus, Peters, Mon. Berl. Ac. 1871, p. 580; Peters, Ann. Mus. Civ. Genova, iii. 1872, p. 44, pl. vi. fig. 3; Boul. Cat. Batr. Sal. p. 99.

A specimen of this elegant little spotted Frog, only previously recorded from Borneo, was caught in the jungle on Bukit Timah, Singapore, in Feb. 1896, by Dr. Hanitsch, of the Raffles Museum.

Hab. Malay Peninsula and Borneo.

17. IXALUS ASPER, Blgr.

Ixalus asper, Boul. P. Z. S. 1886, p. 415, pl. xxxix. fig. 1.

This species was described from specimens sent to the British Museum by Mr. Wray; a pair were "caught breeding in the water on Hill Garden, Larut, Perak, at an elevation of 3300 feet."

Hab. Malay Peninsula, Burma.

Family ENGYSTOMATIDÆ.

18. CALOPHRYNUS PLEUROSTIGMA, Tschudi.

Calophrynus pleurostigma, Boul. Cat. Batr. Sal. p. 158; Boul. Fauna Brit. Ind., Rept. p. 490 (palate fig.).

I obtained one young specimen in the jungle on Bukit Timah, Singapore; this species does not seem to have been previously recorded from the Straits Settlements.

Hab. Burma, South China, Malay Peninsula, Borneo, Natunas.

19. MICROHYLA ACHATINA, Boie.

Microhyla achatina, Boul. Cat. Batr. Sal. p. 166.

There are ♂ and ♀ specimens in the British Museum from Malacca from Mr. Hervey.

Hab. Tenasserim, Malay Peninsula, Sumatra, Java, and Moluccas.

20. MICROHYLA BERDMORII, Blyth.

Microhyla berdmorii, Boul. Catr. Batr. Sal. p. 166.

Boulenger (Fauna Brit. India, Reptiles p. 492) says, "Mr. W. L. Sclater recently communicated to me a specimen obtained by Mr. Davison at Malacca."

Hab. Burma, Camboja, Malay Peninsula.

21. CALLULA PULCHRA, Gray.

Hyladactylus bivittatus, Cantor, p. 143.
Callula pulchra, Boul. Cat. Batr. Sal. p. 170 (hand etc. fig.).

Cantor obtained a male from a field near Malacca; I have not heard of its occurring in Penang.

I have been told by both English and natives that this Frog was unknown in Singapore until some nine or ten years ago, when it was introduced by a half-caste, why it is not known, and that it rapidly spread about the island. It is now well-known as the